PRACTICE
for TAKS
Success

Grade 3

Harcourt
SCHOOL PUBLISHERS

Visit *The Learning Site!*
www.harcourtschool.com

TEXAS HSP Math

ISBN 13: 978-0-15-356902-9
ISBN 10: 0-15-356902-6

3 4 5 6 7 8 9 10 073 16 15 14 13 12 11 10 09 08

CONTENTS

Writing Griddable Responses

Some tests include griddable responses. When you find an answer, write the answer at the top of the grid. Then fill in the matching bubbles below your numbers.

EXAMPLE

George bought 4 shirts for $12 each. How much did George spend on the shirts he bought?

Use multiplication to find the total amount George spent on shirts; $4 \times 12 = 48$. So, George spent $48 on the shirts. Place the numbers in the grid and darken the bubbles that match each number.

© Harcourt

Drawing a Picture

Often, it is easier to find a solution to a question if you draw a picture to show the information in the question.

EXAMPLE

Rectangle A and rectangle B share a side of 12 inches. The shorter side of rectangle A is 6 inches. The length of the longer side of rectangle B is 24 inches. What is the perimeter of rectangle B?

| | | |
| A | 12 in. | B |

6 in. | ←——— 24 in. ———→

(handwritten work) 24 · B.A. 12 · 24 + 48 · 48 +12 · 60 · 72 + 12 · 72 · 60

- ○ 36 in.
- ● 72 in.
- ○ 104 in.
- ○ 288 in.

Drawing the rectangles helps you see which numbers to multiply to find the area of rectangle B.

To find the perimeter, add each of the sides of rectangle B: 24 + 24 + 12 + 12 = 72.

So, the correct answer is 72 inches.

Estimating an Answer

You can use estimation to find an answer, check an answer, or eliminate possible answers in multiple-choice questions.

EXAMPLE

Carter goes school-supply shopping. He has only one bill in his wallet and is able to buy one of everything on his list. He receives only coins for his change back.

School Supply List	
Supply	**Cost**
1 pack of pencils	$2.99
1 pack of notebook paper	$1.59
1 spiral notebook	$1.79
1 binder	$2.69

Which bill could Carter have in his wallet?

- ○ $1 ✗
- ○ $5 ✗
- ● $10 ✓
- ○ $20 ✗

You can estimate the cost of each item and add the estimated numbers mentally. $2.99 rounds up to $3.00, $1.59 rounds up to $2.00, $1.79 rounds up to $2.00, and $2.69 rounds up to $3.00.

$3 + $2 + $2 + $3 = $10 ✓

The first and second answer choice are too low. The fourth answer choice, $20, would allow Carter to buy one of everything; however, he would get bills and coins for change. So, Carter has a $10 bill in his wallet.

TAKS Strategies

v

Practice for TAKS Success

TAKS Strategies

Eliminating Answer Choices

You can solve some math problems without doing actual computation. You can use mental math, estimation, or logical reasoning to help you eliminate answer choices and save time.

EXAMPLE

Which number is the closest estimate for 257 + 410?

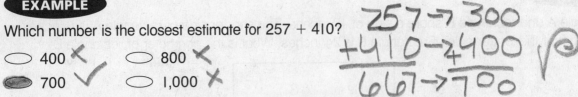

- ⊘ 400 ✗
- ⊘ 800 ✗
- ⬤ 700 ✓
- ⊘ 1,000 ✗

You can use logical reasoning to eliminate the first answer choice, 400, because it is too small. The estimated sum has to be greater than 400 because one of the numbers, 410, rounds to 400 and you are adding 257 and 410.

The fourth answer choice, 1,000, can be eliminated because it is too large. The estimated sum will be less than 1,000.

Round 257 up to 300 and 410 down to 400. Then find the sum of 300 and 400; 300 + 400 = 700. You can eliminate the third answer choice because it is greater than 700.

The second answer choice, 700, is the closest estimate.

Answering the Question Asked

EXAMPLE

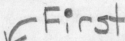

Deisha's class has a spelling bee each week. She and three other students keep track of how many words they have spelled correctly so far this year. Their results are shown in the table.

Correct Spelling Bee Words	
Name	Number of Words
Deisha	28
Mark	30
Lindsey	18
Roshan	24

Correct Spelling Bee Words	
Name	Number of Words
Deisha	○○○○○○○
Mark	○○○○○○○(
Lindsey	
Roshan	○○○○○○

Key: Each ○ = 4 words.

How many circles should be shown in the pictograph next to Lindsey's name?

- ⬤ $4\frac{1}{2}$ ✓
- ⊘ 18 ✗
- ⊘ $7\frac{1}{2}$ ✗
- ⊘ 24 ✗

The question asks for how many circles should go next to Lindsey's name. The second answer choice, $7\frac{1}{2}$, shows how many circles should go next to Mark's name, not Lindsey's name. The third answer choice, 18, shows how many **words** Lindsey spelled correctly, not the number of circles that should be shown next to her name. The last answer choice, 24, is how many words Roshan spelled correctly, not Lindsey. The answer to the question is $4\frac{1}{2}$.

Practice for TAKS Success   vi   **TAKS Strategies**

© Harcourt

Name __Parker chatham__ 7#
2/25/09

1 ⬇ 3.1A In 2000, the population of Martindale, Texas was 953. Which shows 953 written in expanded form? Mark your answer.

- ⬭ 9 + 5 + 3 ✗
- ⬭ 900 + 5 + 3 ✗
- ⬤ 900 + 50 + 3 ✓ ?
- ⬭ 90 + 50 + 30 ✗

2 3.8 Which describes the figure? Mark your answer.

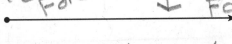

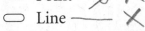

- ⬭ Point
- ⬭ Line
- ⬤ Ray
- ⬭ Line segment

3 ⬇ 3.3B An art exhibit at a gallery in Dallas had a photography section with 5 rows and 9 photographs in each row. Which shows how many total photographs were in the exhibit? Mark your answer.

- ⬤ 45
- ⬭ 59
- ⬭ 62
- ⬭ 93

9 × 5 = 45
5 × 9 = 45

4 ⬇ 3.1B In 2000, the population of Muenster, Texas was 1,556, the population of Hubbard, Texas was 1,586, and the population of Blanco, Texas was 1,505. Which shows the numbers ordered from greatest to least? Mark your answer.

- ⬭ 1,505; 1,556; 1,586 ✗
- ⬤ 1,586; 1,556; 1,505 ✓
- ⬭ 1,505; 1,586; 1,556 ✗
- ⬭ 1,556; 1,505; 1,586 ✗

G → L
1,586
1,586
1,505

1,586
1,556
1,505

© Harcourt

Practice A

1

Practice for TAKS Success

5 ✦ **3.11B**   Anne is putting a border around a picture that she drew. The picture is 4 inches tall and 9 inches wide. Which is the perimeter of the border? Mark your answer.

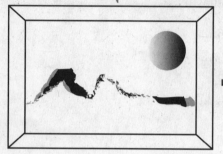

9 inches

4 inches

○ 13 inches
◉ 26 inches
○ 36 inches
○ 94 inches

$$
\begin{array}{r}
4\text{in.} \\
+4\text{in.} \\
\hline
8\text{in.} \\
18\text{in} \\
+8\text{in} \\
\hline
26\text{in}
\end{array}
$$

6 ✦ **3.8**   Which of the following has a bottom, side, or top view that looks like the figure? Mark your answer.

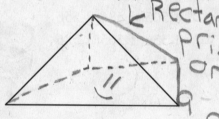

Rectangular prism or a cone

○ Rectangular prism ✓ or ?
◉ Cone ✓
○ Cylinder
○ Sphere

7 ✦ **3.7B; 3.14C**   The table shows the cost of tickets at an amusement park in Dallas, Texas. Which of the following is a rule for the table? Mark your answer.

Tickets	2	4	6	8	10
Cost	$20	$40	$60	$80	$100

○ Multiply the number of tickets by $4.
○ Add $8 to the number of tickets.
◉ Multiply the number of tickets by $10.
○ Add $10 to the number of tickets.

© Harcourt

8 🔸 **3.13B** Nancy made a tally table to record her friends' votes for their favorite color. Which shows how many of Nancy's friends did NOT vote for purple? Mark your answer.

Favorite Color	
Color	Tally
Red	ⅢⅠ
Green	ⅢⅠ ⅢⅠ ⅢⅠ
Blue	ⅢⅠ ⅠⅠ
Purple	ⅢⅠ ⅢⅠ

- ⬯ 10 ✗
- ⬯ 20 ✗
- ⬤ 27 votes ✓
- ⬯ 37 ✗

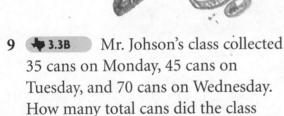

9 🔸 **3.3B** Mr. Johson's class collected 35 cans on Monday, 45 cans on Tuesday, and 70 cans on Wednesday. How many total cans did the class collect? Mark your answer.

- ⬯ 145 ✓ ✗
- ⬤ 150 ✓
- ⬯ 165 ✗
- ⬯ 180 ✗

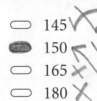

$$35$$
$$+45$$
$$70$$
$$150$$

10 🔸 **3.11A; 3.14A** Jason wants to know the length of his kitchen. Which is the best estimate of the measure for the room? Mark your answer.

- ⬯ 4 feet ✗
- ⬤ 4 yards maybe?
- ⬯ 40 feet or
- ⬯ 400 yards maybe

11 🔸 **3.13C** Which is the probability of landing on a blue section of the spinner if all the sections are equal size? Mark your answer.

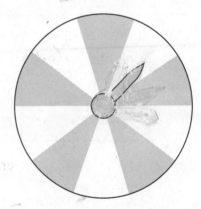

- ⬤ Impossible
- ⬯ Unlikely ✓
- ⬯ Likely or ✓
- ⬯ Certain ✓

12 3.9C; 3.14C Which of the following figures appears to have a line of symmetry? Mark your answer.

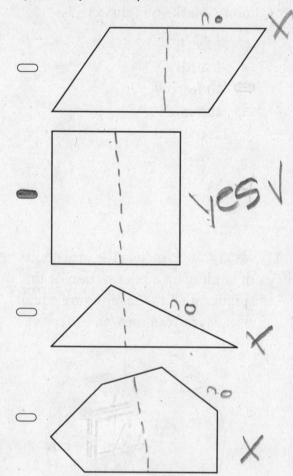

13 3.6A Which shows the missing figure? Mark your answer.

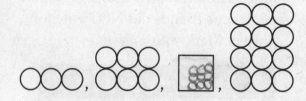

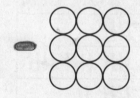

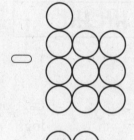

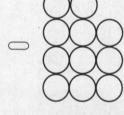

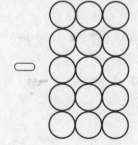

14 3.6A Cody keeps track of the number of daily visitors that attend the Austin Museum of Art in Austin, Texas. How many people would attend the museum on the next day, if the pattern continues? Mark your answer.

445, 450, 455, 460, 465, ▢

- ⬭ 466
- ⬬ 470
- ⬭ 475
- ⬭ 500

15 3.4C A group of students orders 15 hot dogs at a restaurant in Lubbock, Texas. Each student ate 3 hot dogs. Which shows how many students were in the group? Mark your answer.

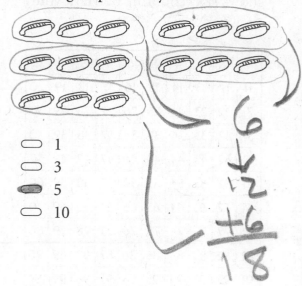

- ⬭ 1
- ⬭ 3
- ⬬ 5
- ⬭ 10

16 3.5B The population of Chillicothe, Texas is 798. The population of Krugerville, Texas is 908. About how much greater is the population of Krugerville? Mark your answer.

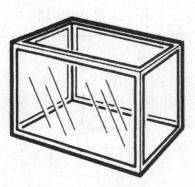

- ⬬ about 100
- ⬭ about 200
- ⬭ about 300
- ⬭ about 400

17 3.11E Jessica wants to find the capacity of a large aquarium. Which unit should she use to measure? Mark your answer.

- ⬭ cup
- ⬭ pint
- ⬭ quart
- ⬬ gallon

18 🔻 3.8 How are the figures alike? Mark your answer.

not the faces

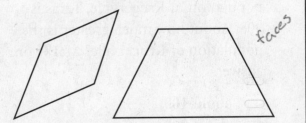

- ⬭ They both have equal size angles.
- ⬭ They both have 8 sides.
- ⬭ They both have curved lines.
- ⬬ They both have 4 sides.

19 🔻 3.5A The driving distance from Austin to San Antonio is about 79 miles. Which shows 79 rounded to the nearest ten? *80*

Record your answer in the boxes below. Then fill in the bubbles. Be sure to use the correct place value.

8	0
⓪	⓪
①	①
②	②
③	③
④	④
⑤	⑤
⑥	⑥
⑦	⑦
⑧	⑧
⑨	⑨

20 🔻 3.4A Which multiplication sentence is modeled by the array? Mark your answer.

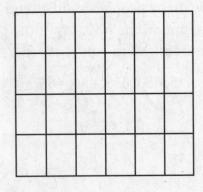

- ⬭ $4 \times 3 = 12$
- ⬭ $4 \times 4 = 16$
- ⬬ $4 \times 6 = 24$
- ⬭ $6 \times 6 = 36$

21 🔻 3.6A; 3.16A Use the hundred chart. Which of the following set of numbers shows skip counting 6? Mark your answer.

1	2	3	4	5	6	7	8	9	10
11	12	13	14	15	16	17	18	19	20
21	22	23	24	25	26	27	28	29	30
31	32	33	34	35	36	37	38	39	40
41	42	43	44	45	46	47	48	49	50
51	52	53	54	55	56	57	58	59	60
61	62	63	64	65	66	67	68	69	70
71	72	73	74	75	76	77	78	79	80
81	82	83	84	85	86	87	88	89	90
91	92	93	94	95	96	97	98	99	100

- ⬭ 16, 26, 32
- ⬬ 12, 18, 24
- ⬭ 14, 16, 20
- ⬭ 22, 26, 29

22 **3.13C** What is the probability of pulling a tile marked "O" out of the bag if the tiles are the same size and shape? Mark your answer.

- ⬭ Impossible
- ⬭ Unlikely
- ⬭ Likely
- ⬭ Certain

23 **3.7B; 3.14C** The table shows the cost of sandwiches at the café near Juan's house. Find the cost of 12 sandwiches. Mark your answer.

Sandwiches	2	4	6	8	10	12
Cost	$10	$20	$30	$40	$50	☐

$5 × 6
$60

- ⬭ $51
- ⬭ $52
- ⬭ $54
- ⬭ $60

24 **3.11D** Sonya has a kitten that weighs 64 ounces. Which shows how many pounds are equal to 64 ounces? Mark your answer.

- ⬭ 2 pounds
- ⬭ 4 pounds
- ⬭ 5 pounds
- ⬭ 6 pounds

25 **3.11F** Sebastian has a box in the shape of the figure below. Which is the volume of the box? Mark your answer.

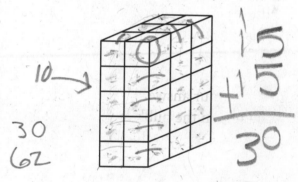

10 →
30
62

- ⬭ 30 cubic units
- ⬭ 52 cubic units
- ⬭ 75 cubic units
- ⬭ 100 cubic units

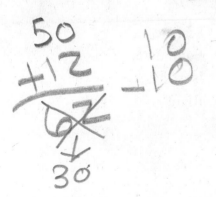

7

Name _____

26 🔻 **3.13A** Stuart made a pictograph to show how many fruits he has. Which shows how many apples Stuart has? Mark your answer.

Types of Fruit	
Apples	🥣 🥣
Oranges	🥣
Bananas	🥣 🥣
Pears	🥣
Each 🥣 **= 5 fruits**	

- ⬭ 5
- ⬭ 10
- ⬭ 15
- ⬭ 20

27 🔻 **3.8** Which correctly names this triangle? Mark your answer.

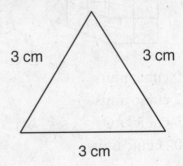

3 cm, 3 cm, 3 cm

- ⬭ Scalene
- ⬭ Equilateral
- ⬭ Isosceles
- ⬭ Right

28 🔻 **3.13A; 3.14C** Which scale would be best to use to make a bar graph of the data shown in the pictograph?

Favorite Music	
Rock	★★
Country	★★★⯪
Hip-hop	★★★★
Classical	★
Each Star = 5 votes	

- ⬭ 0, 5, 10, 15, 20, 25, 30
- ⬭ 0, 2, 4, 6, 8, 10, 12
- ⬭ 0, 10, 20, 30, 40, 50, 60
- ⬭ 0, 8, 16, 24, 32, 40, 48

29 ★3.11A Which shows the length of the pen to the nearest half-inch? Mark your answer.

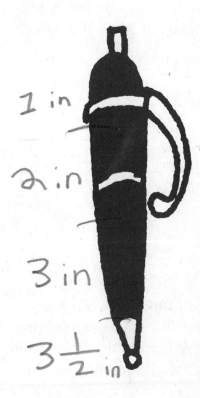

⬭ $3\frac{1}{2}$ inches
⬭ 4 inches
⬭ $4\frac{1}{2}$ inches
⬭ 5 inches

30 ★3.9C Which figure would complete the given letter so that the line is a line of symmetry? Mark your answer.

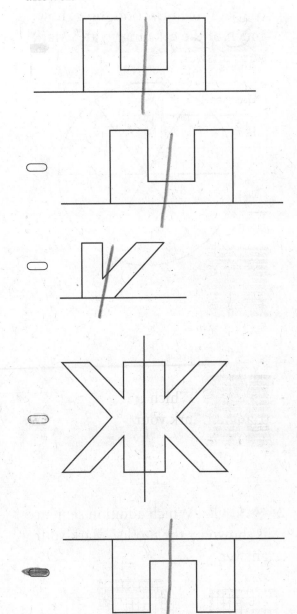

31 ✦**3.2B; 3.14B** Maria and Ingrid bake a cake. The cake is cut into 6 equal pieces. Maria eats $\frac{1}{3}$ of the pieces. Ingrid eats a greater amount than Maria. Which fraction shows how much of the cake Ingrid ate? Mark your answer.

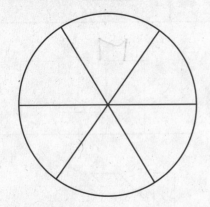

- ⬭ $\frac{1}{6}$
- ⬭ $\frac{2}{6}$
- ⬭ $\frac{1}{4}$
- ⬤ $\frac{1}{2}$

32 ✦**3.3A** Which addition sentence is shown by the model? Mark your answer.

 +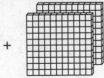

- ⬭ $100 + 130 = 230$
- ⬭ $100 + 236 = 436$
- ⬤ $100 + 236 = 336$
- ⬭ $100 + 200 = 300$

$$\begin{array}{r} 100 \\ +236 \\ \hline 336 \end{array}$$

33 ✦**3.1C; 3.14C** Jose bought a comic book. He paid with a ten-dollar bill and received a five-dollar bill and 5 pennies in change. Which is the amount the comic book cost? Mark your answer.

- ⬭ $2.95
- ⬭ $3.95
- ⬭ $4.95
- ⬤ $5.05

34 ✦**3.6A** Lauren drew a pattern. The pattern is: triangle, square, circle, pentagon. What will the 9th and 10th figures be? Mark your answer.

- ⬭ Triangle, square
- ⬭ Square, circle
- ⬭ Circle, pentagon
- ⬤ Pentagon, triangle

35 **3.11C** Kyoko draws a rectangle with two sides that measure 12 inches and two sides that measure 4 inches. Which is the area of the rectangle? Mark your answer.

12 inches

4 inches

○ area = 28 square inches
○ area = 32 square inches
● area = 48 square inches ?
○ area = 40 square inches

36 **3.6A** Which shows the next two shapes in the pattern?

○△○○△○○○△○○○

○ △, △

● △, ○

○ ○, △

○ ○, ○

37 **3.9A** Which figure is congruent to the given figure? Mark your answer.

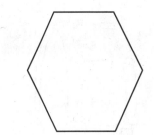

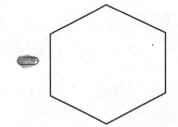

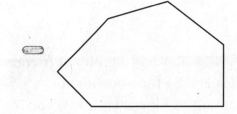

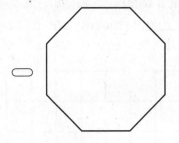

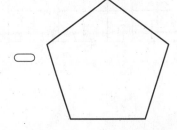

38 3.8   Angela likes to play her father's drum. Which best describes the shape of the drum? Mark your answer.

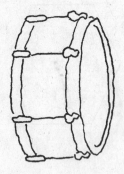

○ Triangular prism
● Cylinder
○ Cube
○ Rectangular prism

39 3.4C   Ray and a group of friends each spend $4 for a burrito at a restaurant in Muenster, Texas. The group spent a total of $40. Which shows how many students are in the group? Mark your answer.

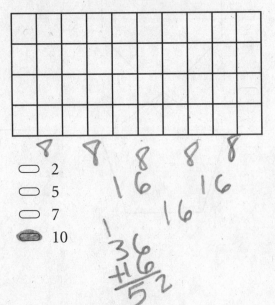

○ 2
○ 5
○ 7
● 10

40 3.13C   Charlie is going to pull one marble out of the bag. What is the probability of pulling a white marble out of the bag if the marbles are all the same size? Mark your answer.

○ Impossible
○ Unlikely
○ Likely
● Certain

© Harcourt

1 🔽 **3.10** Which number is less than point *Y* on the number line? Mark your answer.

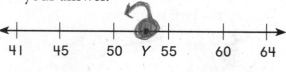

- ⬭ 50
- ⬭ 55
- ⬭ 56
- ⬭ 60

2 🔽 **3.3B** The Museum of Fine Arts in Houston had 354 visitors on Saturday and 320 visitors on Sunday. How many total visitors did the museum have over the two days? Mark your answer.

- ⬭ 444
- ⬭ 674
- ⬭ 677
- ⬭ 774

$$\begin{array}{r} 354 \\ +\ 320 \\ \hline 674 \end{array}$$

3 🔽 **3.12B** Ethan meets a friend at the park at the time shown on the clock. Which time does the clock show? Mark your answer.

- ⬭ 1:03
- ⬭ 2:02
- ⬭ 2:03
- ⬭ 3:02

4 🔽 **3.1A** Texas covers a total area of 268,581 square miles. Which is the value of the underlined digit? Mark your answer.

- ⬭ 8
- ⬭ 80
- ⬭ 800
- ⬭ 8,000

5 **3.3A** Which addition sentence is shown by the model? Mark your answer.

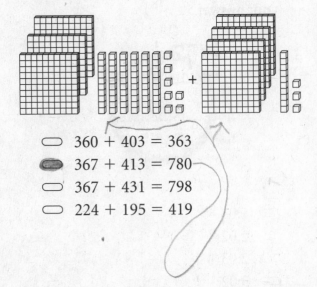

- ⬭ $360 + 403 = 363$
- ⬬ $367 + 413 = 780$
- ⬭ $367 + 431 = 798$
- ⬭ $224 + 195 = 419$

6 **3.1A** The driving distance from Houston, Texas to Seattle, Washington is about 2,430 miles. Which shows 2,430 miles written in expanded form? Mark your answer.

- ⬭ $2,000 + 100 + 10$
- ⬭ $2,000 + 200 + 30$
- ⬬ $2,000 + 400 + 30$
- ⬭ $4,000 + 400 + 30$

7 **3.1B** The Renaissance Tower is 886 feet tall, and the Williams Tower is 901 feet tall. Which statement is correct? Mark your answer.

- ⬭ $886 = 901$
- ⬭ $901 < 886$
- ⬭ $886 > 901$
- ⬬ $901 > 886$

8 **3.1C** Which is the missing number? Mark your answer.

2 quarters + ☐ nickels = $1.00

- ⬭ 5
- ⬬ 10
- ⬭ 20
- ⬭ 30

© Harcourt

9 3.5A Lincoln Elementary School has 208 second-grade students and 222 third-grade students. About how many second and third-grade students are there. Mark your answer.

○ About 400
○ About 440
○ About 470
○ About 500

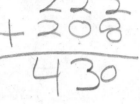

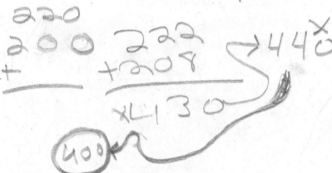

10 3.1C Which amount is the greatest? Mark your answer.

○ 5 quarters, 10 nickels
○ 5 quarters, 10 dimes
○ 5 quarters, 9 dimes
○ 5 quarters, 12 nickels

11 3.12B Lisa ran in the Austin Marathon. Her total time is shown on the clock below. Which was her time for the race? Mark your answer.

○ 8 hours, 48 minutes, 18 seconds
○ 3 hours, 02 minutes, 42 seconds
○ 58 hours, 4 minutes, 18 seconds
○ 4 hours, 58 minutes, 18 seconds

12 3.3A Which number sentence is in the same fact family for the numbers 58, 90, 32? Mark your answer.

○ 58 − 32 = 26
○ 58 + 90 = 148
○ 90 − 58 = 32
○ 90 + 32 = 122

13 3.1B Lakewood Elementary School had 287 students in 2000, 264 students in 2001, and 270 students in 2003. Which shows the numbers ordered from greatest to least? Mark your answer.

○ 267; 264; 270
○ 287; 264; 270
○ 287; 270; 264
○ 270; 287; 264

14 3.1A The NASA Space shuttle orbits 46,000 feet above the earth. Which is the value of the underlined digit? Mark your answer.

○ 6
○ 60
○ 600
● 6,000

15 3.3B Nelson has 245 stamps in his collection. Tracy has 377 stamps. Which shows how many more stamps Tracy has in her collection? Mark your answer.

○ 127
● 132
○ 142
○ 153

16 3.5A Which shows the number shown by point Z rounded to the nearest ten? Mark your answer.

○ 310
● 320
○ 323
○ 400

17 3.1C Which is the correct amount shown? Mark your answer.

○ $8.35
○ $9.25
○ $9.35
● $10.35

18 3.5B Use compatible numbers to estimate the difference between 868 and 442. Mark your answer.

○ 860 − 450 = 410
○ 860 − 440 = 420
● 870 − 440 = 430
○ 880 − 440 = 440

© Harcourt

19 3.3B The Heard Natural Science Museum and Wildlife Sanctuary had 856 visitors on Saturday. The museum had 745 visitors on Sunday. How many total visitors did the museum have over the two days? Mark your answer.

- 1,600
- 1,601
- 1,800
- 1,801

20 3.1B In 2000, the population of Perryton, Texas was 7,774, the population of Hilgado, Texas was 7,732, and the population of Glen Heights, Texas was 7,224. Which shows the numbers ordered from least to greatest? Mark your answer.

- 7,224; 7,732; 7,774
- 7,732; 7,774; 7,224
- 7,224; 7,774; 7,732
- 7,732; 7,224; 7,774

21 3.3A Which number sentence is in the same fact family for the numbers 31, 77, 46? Mark your answer.

- 77 + 31 = 108
- 46 − 31 = 15
- 46 + 31 = 77
- 77 + 46 = 123

22 3.3A Which number makes this number sentence true? Mark your answer.

$$267 + \square = 300$$

- 7
- 10
- 24
- 33

23 3.1A Which is the value of the underlined digit in 50,147? Mark your answer.

- 5
- 500
- 5,000
- 50,000

24 3.5B Use compatible numbers to estimate the sum of 549 and 751. Mark your answer.

- 550 + 750 = 1,300
- 540 + 750 = 1,290
- 500 + 700 = 1,200
- 550 + 700 = 1,250

25 3.3A Which number makes this number sentence true? Mark your answer.

$$45 + \square = 63$$

- 17
- 18
- 19
- 20

26 ★ 3.5A   The driving distance from El Paso to Houston is about 695 miles. Which shows 695 rounded to the nearest hundred? Mark your answer.

- ⬭ 600
- ⬬ 700
- ⬭ 800
- ⬭ 900

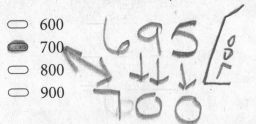

27 ★ 3.12B   Which time is half past five? Mark your answer.

- ⬭ 4:30
- ⬭ 5:00
- ⬬ 5:30
- ⬭ 6:30

28 ★ 3.1C   Which amount is least? Mark your answer.

- ⬬ 9 nickels, 34 pennies
- ⬭ 7 nickels, 58 pennies
- ⬭ 2 quarters, 7 nickels
- ⬭ 3 quarters, 9 pennies

29 ★ 3.3A   Which subtraction sentence is shown by the model? Mark your answer.

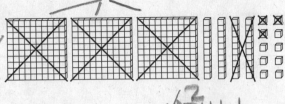

- ⬬ 350 − 323 = 27
- ⬭ 350 − 325 = 25
- ⬭ 350 − 322 = 28
- ⬭ 350 − 323 = 23

30 ★ 3.5B   Use compatible numbers to estimate the difference between 438 and 362. Mark your answer.

- ⬬ 440 − 360 = 80
- ⬭ 400 − 300 = 100
- ⬭ 430 − 360 = 800
- ⬭ 440 − 370 = 810

31 ★ 3.3A   Which addition sentence is shown by the model? Mark your answer.

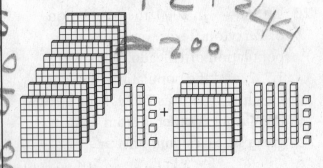

- ⬬ 724 + 244 = 968
- ⬭ 722 + 222 = 944
- ⬭ 724 + 224 = 948
- ⬭ 722 + 224 = 926

32 ★ 3.1C   Jen bought a ticket for admission to a local amusement park. The ticket cost $5.95. Jen paid for the ticket with a ten-dollar bill. Which is the amount of change Jen received? Mark your answer.

- ⬭ $3.05
- ⬬ $4.05
- ⬭ $4.15
- ⬭ $4.95

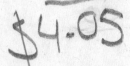

© Harcourt

1 🔶 **3.6A** Which shows the next two figures in the pattern?

○△○▭○△○▭○▭○△○__ __

C+CRC+CRC+C

- ○△
- ▭△
- ○▭
- ▭○

2 🔶 **3.4A** Dan and his family are going on a camping trip in Davis Mountain State Park. The park charges an $11 fee per day of camping. Which multiplication sentence shows the total camping fees for 5 days? Mark your answer.

Fri.	Sat.	Sun.	Mon.	Tues.
$11	$11	$11	$11	$11

- $10 × 5 = $50
- $5 × 5 = $25
- $11 × 5 = $55
- $11 × 11 = $121

22
22
+11
55

3 🔶 **3.4A** Bonnie is making a design with 9 squares. Which multiplication sentence shows how many sides 9 squares have? Mark your answer.

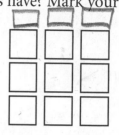

- 4 × 3 = 12
- 4 × 4 = 16
- 4 × 9 = 36   *not given*
- 9 × 9 = 81

4 🔶 **3.1C** Which is the amount shown? Mark your answer.

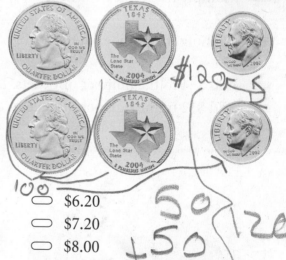

- $6.20
- $7.20
- $8.00
- $8.20

$8.20

© Harcourt

5 **3.1A** Julie reads a book that has 427 pages. Which is the value of the underlined digit? Mark your answer.

⬭ 4

⬭ 40

⬬ 400

⬭ 4,000

6 **3.7B** The table shows the cost of tickets at the Dallas Museum of Art. Which is the rule for the table? Mark your answer.

Tickets	2	3	4	9	11
Cost	$24	$36	$48	$108	$132

⬭ Multiply the number of tickets by $12.

⬭ Multiply the number of tickets by $24.

⬭ Add $24 to the number of tickets.

⬬ Add $12 to the number of tickets.

7 **3.4A** Charles has 6 sets of paints. Each set has 5 tubes of different colors. Which shows how many total tubes Charles has? Mark your answer.

⬭ 11

⬭ 25

⬬ 30

⬭ 35

8 **3.5B** Use compatible numbers to estimate the difference between 578 and 223. Mark your answer.

⬬ 575 − 225 = 350

⬭ 600 − 300 = 300

⬭ 575 − 250 = 325

⬭ 550 − 250 = 300

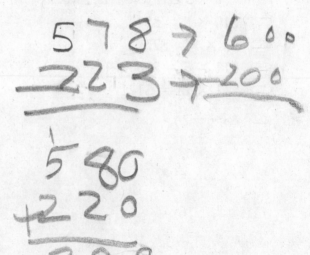

© Harcourt

9 3.13B The bar graph shows the results of the survey of the students' favorite school subjects. Which shows how many total votes English and Science received? Mark your answer.

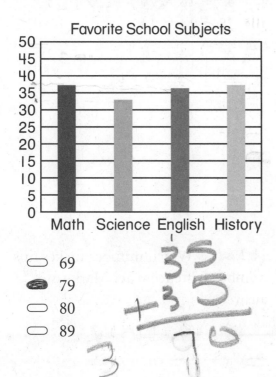

Favorite School Subjects

- ⊂⊃ 69
- ⊂●⊃ 79
- ⊂⊃ 80
- ⊂⊃ 89

10 3.1B Matthew bicycles 67 miles the first week, 48 miles the second week, and 74 miles the third week. Which shows the numbers ordered from least to greatest? Mark your answer.

- ⊂⊃ 48; 74; 67
- ⊂⊃ 74; 48; 67
- ⊂●⊃ 67; 48; 74
- ⊂⊃ 48; 67; 74

11 3.6A Elsie drew this pattern. The pattern unit is:

What will be the 11th and 12th figures? Mark your answer.

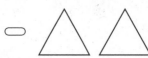

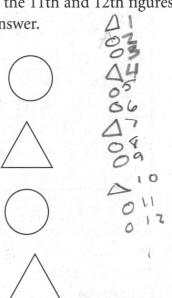

12 3.7B The table shows the cost of magazines at a local bookstore. Which is the cost of 8 magazines? Mark your answer.

Magazines	4	5	6	7	8
Cost	$12	$15	$18	$21	☐

- ⊂⊃ $16
- ⊂⊃ $18
- ⊂●⊃ $24
- ⊂⊃ $30

13 🔻 **3.6A** Use the hundreds chart. Which of the following set of numbers shows skip counting by 11?

1	2	3	4	5	6	7	8	9	10
11	12	13	14	15	16	17	18	19	20
21	22	23	24	25	26	27	28	29	30
31	32	33	34	35	36	37	38	39	40
41	42	43	44	45	46	47	48	49	50
51	52	53	54	55	56	57	58	59	60
61	62	63	64	65	66	67	68	69	70
71	72	73	74	75	76	77	78	79	80
81	82	83	84	85	86	87	88	89	90
91	92	93	94	95	96	97	98	99	100

- 33, 55, 66
- 21, 32, 43
- 20, 24, 28
- 49, 58, 67

14 🔻 **3.13A** Which scale would be best to use to make a bar graph of the data shown in the pictograph?

Favorite Exercise	
Swimming	★ ★
Yoga	★ ⭐
Dancing	★ ★ ★
Bicycling	★ ★ ⭐

Key: Each ★ = 12 Votes

- 0, 5, 10, 15, 20, 25, 30
- 0, 2, 4, 6, 8, 10, 12
- 0, 10, 20, 30, 40, 50, 60
- 0, 12, 24, 36, 48, 60

15 🔻 **3.3A** Which number sentence is in the same fact family for the numbers 14, 42, 28? Mark your answer.

- 42 − 14 = 28
- 42 + 28 = 70
- 28 − 14 = 14
- 42 + 14 = 56

16 🔻 **3.3A** Which number makes this number sentence true? Mark your answer.

$$452 - \square = 187$$

- 132
- 255
- 265
- 639

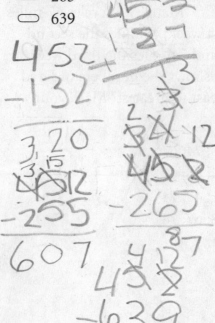

© Harcourt

17 3.4A George's Bakery makes 10 dozen biscuits each day. There are 12 biscuits in a dozen. Which shows how many total biscuits the bakery makes each day? Mark your answer.

⊖ 22
⊖ 112
⬤ 120
⊖ 1200

18 3.13B Sam made a tally table to record his friends' votes for their favorite sports team. Which shows how many of Sam's friends voted for the Cowboys? Mark your answer.

Favorite Sports Team	
Team	Tally
Mavericks	卌
Astros	卌 卌 III
Cowboys	卌 卌 卌 I
Longhorns	卌 卌

⊖ 14
⊖ 15
⬤ 16
⊖ 21

19 3.4A Which multiplication sentence does the array show? Mark your answer.

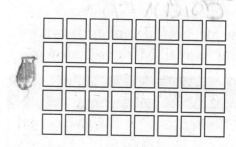

⊖ 5 + 8 = 13
⊖ 5 × 5 = 25
⊖ 8 × 8 = 64
⬤ 5 × 8 = 40

20 3.3B Mary's school has 238 second-grade students and 311 third-grade students. What is the total number of second and third-grade students in Mary's school? Mark your answer.

⊖ 73
⊖ 539
⬤ 549
⊖ 649

21 ○ **3.13C** Keisha made a tally table to record her friends' votes for their favorite Texas animals. Which shows how many of Keisha's friends did NOT vote for mockingbird as their favorite Texas animal? Mark your answer.

Favorite Animal	
Subject	**Tally**
Longhorn	ⅢⅢ ⅢⅢ III
Mockingbird	ⅢⅢ ⅢⅢ ⅢⅢ III
Bat	ⅢⅢ II
Armadillo	ⅢⅢ ⅢⅢ ⅢⅢ I

○ 18
○ 23
⊖ 36
○ 54

22 ○ **3.13B** The bar graph shows the results of Mathew's survey of favorite Texas flowers. Which flower received more votes than Indian Blanket? Mark your answer.

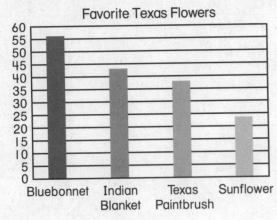

Favorite Texas Flowers

○ Texas Paintbrush
⊖ Bluebonnet
○ Indian Blanket
○ Sunflower

23 ○ **3.1A** Jessica drives 1,603 miles on a trip around Texas. Which shows 1,603 written in expanded form? Mark your answer.

○ 1,000 + 300 + 6
⊖ 1,000 + 600 + 3
○ 1,000 + 600 + 30
○ 1 + 6 + 3

24 ○ **3.6A** Use the hundred chart. Which of the following set of numbers shows skip counting 7? Mark your answer.

1	2	3	4	5	6	7	8	9	10
11	12	13	14	15	16	17	18	19	20
21	22	23	24	25	26	27	28	29	30
31	32	33	34	35	36	37	38	39	40
41	42	43	44	45	46	47	48	49	50
51	52	53	54	55	56	57	58	59	60
61	62	63	64	65	66	67	68	69	70
71	72	73	74	75	76	77	78	79	80
81	82	83	84	85	86	87	88	89	90
91	92	93	94	95	96	97	98	99	100

○ 36, 42, 48
⊖ 42, 49, 63
○ 36, 45, 54
○ 12, 24, 36

© Harcourt

25 🔹 **3.6A** Betty's train leaves at 8:00 A.M. The train stops at 8:06 A.M., then at 8:12 A.M., and then at 8:18 A.M. All 7 of the stops are the same distance apart. What time does Betty arrive at the last stop? Mark your answer.

Stops	1	2	3	4	5	6	7
Times	8:06	8:12	8:18	8:24	8:30	8:36	☐

8:42

- ⬭ 8:24 A.M.
- ⬭ 8:36 A.M.
- ⬭ 8:40 A.M.
- ⬬ 8:42 A.M.

26 🔹 **3.5A** Which shows the number shown by point Q rounded to the nearest hundred? Mark your answer.

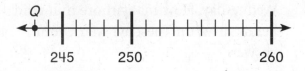

245   250   260

- ⬬ 200
- ⬬ 240
- ⬭ 250
- ⬭ 300

27 🔹 **3.12B** Steve meets a friend at the museum at the time shown on the clock. Which time does the clock show? Mark your answer.

- ⬬ 3:15
- ⬭ 7:45
- ⬭ 8:50
- ⬭ 9:35

28 🔹 **3.4A** Which multiplication sentence is related to the addition sentence? Mark your answer.

$$6 + 6 + 6 + 6$$

- ⬭ $4 \times 4 = 16$
- ⬭ $6 \times 6 = 36$
- ⬬ $6 \times 4 = 24$
- ⬭ $3 \times 6 = 12$

29 3.4A Which multiplication sentence does the array show? Mark your answer.

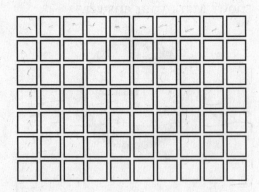

- 7 × 10 = 70
- 8 × 10 = 80
- 9 × 10 = 90
- 10 × 10 = 100

30 3.6A Gisele drew a pattern. The pattern is: pentagon, pentagon, square, triangle. Which will be the 15th and 16th figures? Mark your answer.

- Pentagon, pentagon
- Pentagon, square
- Square, triangle
- Triangle, pentagon

31 3.4A Jack walks 3 miles each day for 4 days while on vacation. Which shows how many total miles Jake walks? Mark your answer.

Friday	Saturday	Sunday	Monday
3 miles	3 miles	3 miles	3 miles

- 5
- 7
- 8
- 12

4 × 3 =

32 3.3B Penny drives 364 miles on Tuesday. She drives 279 miles on Wednesday. How many more miles did Penny drive on Tuesday? Mark your answer.

- 15
- 75
- 85
- 643

1 ★ 3.4C   A group of 18 birds gathered around 2 birdbaths. Each birdbath had the same amount of birds. Which shows how many birds were at each birdbath? Mark your answer.

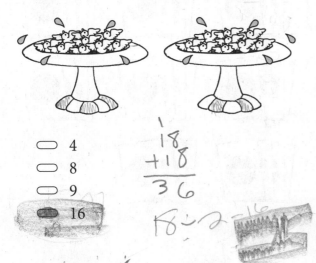

- ⬭ 4
- ⬭ 8
- ⬭ 9
- ⬬ 16

$$\begin{array}{r} 18 \\ +18 \\ \hline 36 \end{array}$$

$18 \div 2 = 16$

2 ★ 3.4C   Jan has 48 blocks to build 6 houses. Each house will have an equal number of blocks. Which shows how many blocks are in each house? Mark your answer.

192
48
48

12

$$\begin{array}{r} 48 \\ +48 \\ \hline 96 \end{array}$$

$$\begin{array}{r} 96 \\ +96 \\ \hline 192 \end{array}$$

$$\begin{array}{r} 192 \\ +48 \\ \hline 240 \end{array}$$

- ⬭ 5
- ⬭ 6
- ⬭ 7
- ⬬ 8

3 ★ 3.4A   Jesse reads 12 comic books a month for 1 month. Which shows how many total comic books Jesse read? Mark your answer.

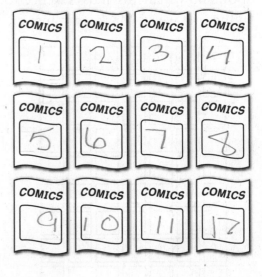

- ⬭ 0
- ⬭ 1
- ⬬ 12
- ⬭ 24

4 ★ 3.8   How many edges does the figure have? Mark your answer.

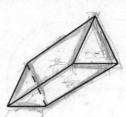

- ⬬ 8
- ⬬ 9
- ⬭ 11
- ⬭ 12

5 3.4C Renee bicycles 70 miles in her first week of training for the Cook Children's Cross in Fort Worth, Texas. She bicycles 10 miles each day. Which shows how many total days Renee rode in the first week? Mark your answer.

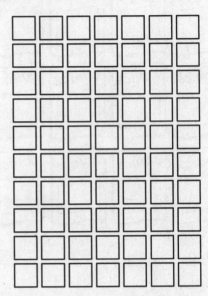

○ 6
⬤ 7
○ 10
○ 17

6 3.8 Which solid can **NOT** have a bottom, side, or top view that looks like the figure? Mark your answer.

○ Square pyramid
⬤ Sphere
○ Cone
○ Cylinder

7 3.4C Janice has 32 trading cards. The cards come in packages of 8 cards each. Which shows how many packages Janice has? Mark your answer.

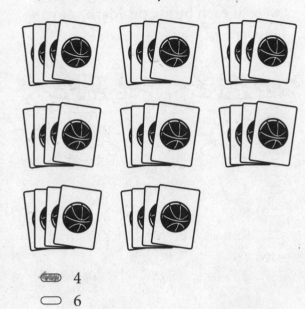

⬤ 4
○ 6
○ 22
○ 24

8 3.1B Which number is less than point B but greater than 80 on the number line? Mark your answer.

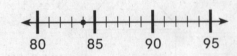

○ 93
○ 91
○ 87
⬤ 82

9 **3.8** Which shows how a square pyramid and a cube are alike? Mark your answer.

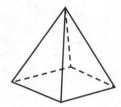 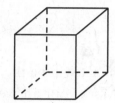

○ Both have a bottom view that is a square.

○ Both have 5 faces.

○ Both have a side view that is a triangle.

○ Both have only 1 vertex.

10 **3.13B** The bar graph shows the results of the survey of students. The students were asked to choose their favorite game. Which two games received a total of 71 votes? Mark your answer.

40
31
——
71

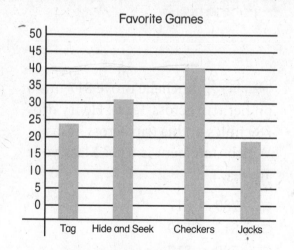

○ Tag and Checkers

○ Hide and Seek and Tag

○ Checkers and Hide and Seek

○ Jacks and Tag

11 **3.9C** Which figure does **NOT** appear to have a line of symmetry? Mark your answer.

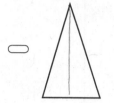

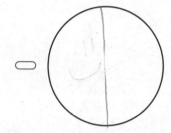

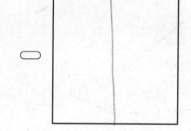

12 3.8 Which figure appears to be a polygon? Mark your answer.

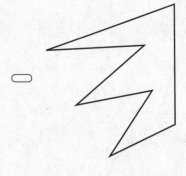

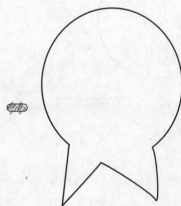

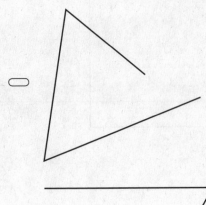

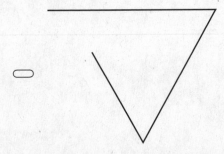

13 3.6A Devon drew a pattern. The pattern unit is:

What will be the 11th and 12th figures? Mark your answer.

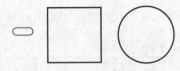

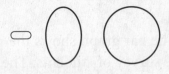

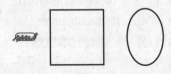

14 3.1B Michelle drove 817 miles on her trip across Texas. Henry drove 789 miles on his trip across Texas. Which statement is true? Mark your answer.

○ 789 > 817
○ 817 < 789
● 789 < 817
○ 817 = 789

15 **3.13B** Lily made a tally table to record her friend's votes for their favorite sports. Which shows how many of Lily's friends voted for basketball? Mark your answer.

Favorite Sport	
Sport	**Tally**
Baseball	ЖЖ II
Football	ЖЖ ЖЖ II
Basketball	ЖЖ ЖЖ ЖЖ IIII
Soccer	ЖЖ ЖЖ III

- ⬭ 7
- ⬭ 12
- ⬭ 19
- ⬭ 51

16 **3.7B** The table shows the cost of packages of trading cards at a local store. Which shows the cost of 5 packages. Mark your answer.

Packages	2	4	5	6	8
Cost	$14	$28	☐	$42	$56

- ⬭ $7
- ⬭ $35
- ⬭ $77
- ⬭ $84

17 **3.9C** Which figure would complete the given letter so that the line is a line of symmetry? Mark your answer.

- ⬭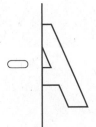
- ⬭
- ⬭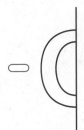
- ⬭

© Harcourt

18 🔻 **3.9A** Which figure appears to be congruent to the given figure shown? Mark your answer.

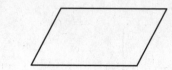

◯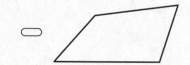

�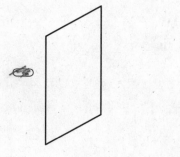

◯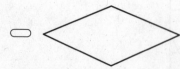

◯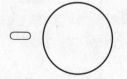

19 🔻 **3.8** Which solid can **NOT** have a bottom, side, or top view that looks like the figure? Mark your answer.

⬭ sphere
◯ square pyramid
◯ cube
◯ rectangular prism

20 🔻 **3.6C** Which number sentence is **NOT** included in the same fact family? Mark your answer.

$$3 \times 8 = 24$$

◯ $8 \times 3 = 24$
◯ $24 \div 8 = 3$
⬛ $24 \times 3 = 72$
◯ $24 \div 3 = 8$

21 🔻 **3.6A** Which of the following set of numbers shows multiples of 9?

1	2	3	4	5	6	7	8	9	10
11	12	13	14	15	16	17	18	19	20
21	22	23	24	25	26	27	28	29	30
31	32	33	34	35	36	37	38	39	40
41	42	43	44	45	46	47	48	49	50
51	52	53	54	55	56	57	58	59	60
61	62	63	64	65	66	67	68	69	70
71	72	73	74	75	76	77	78	79	80
81	82	83	84	85	86	87	88	89	90
91	92	93	94	95	96	97	98	99	100

◯ 9, 27, 46
◯ 18, 32, 48
⬛ 27, 45, 63
◯ 63, 72, 84

22 🔹 **3.4C** Jimmy has 21 plants. He wants to plant them in 7 different pots. Each pot will have the same number of plants. Which shows how many plants are in each pot? Mark your answer.

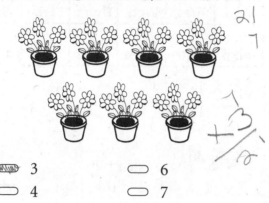

⬛ 3   ⬭ 6
⬭ 4   ⬭ 7

23 🔹 **3.8** Which quadrilateral appears to have 4 square corners? Mark your answer.

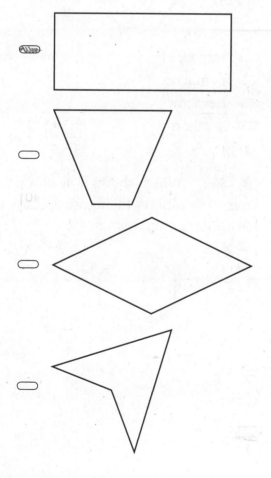

24 🔹 **3.6C** Which number sentence is **NOT** included in the same fact family? Mark your answer.

$$56 \div 8 = 7$$

⬭ $8 \times 7 = 56$
⬭ $56 \div 7 = 8$
⬛ $64 \div 8 = 8$
⬭ $7 \times 8 = 56$

25 🔹 **3.1A** Which is the value of the underlined digit in 2̲87,004? Mark your answer.

⬭ 20
⬭ 200
⬭ 20,000
⬛ 200,000

26 🔹 **3.4A** Which multiplication sentence is modeled by the array? Mark your answer.

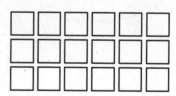

⬭ $3 + 6 = 9$
⬭ $3 \times 3 = 9$
⬛ $3 \times 6 = 18$
⬭ $6 \times 6 = 36$

27 3.4C A group of students are performing 4 different plays. There are an equal number of students in each play and 36 students total. Which shows how many students are in each play? Mark your answer.

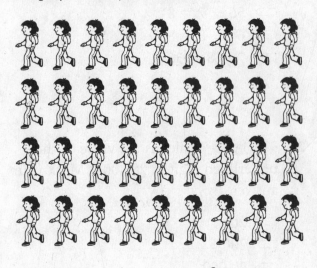

○ 6   ⬭ 9
○ 7   ○ 24

28 3.8 Which shows how many vertices the figure has? Mark your answer.

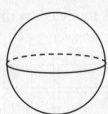

⬭ 0   ○ 2
○ 1   ○ 100

29 3.3A Which is the missing addend? Mark your answer.

$$\square + 9 = 17$$

○ 5   ○ 7
○ 6   ⬭ 8

30 3.4A Christy walks 5 miles each day for 6 days while on vacation. Which shows how many total miles Christy walks? Mark your answer.

Mon.	Tues.	Wed.	Thurs.	Fri.	Sat.
5 miles	5 miles	5 miles	5 miles	5 miles	5 miles

○ 1
○ 11
○ 25
⬭ 30

31 3.8 Which polygon is shown? Mark your answer.

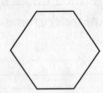

○ trapezoid
○ pentagon
⬭ hexagon
○ octagon

32 3.9C Which shows how many lines of symmetry the figure appears to have? Mark your answer.

○ 3
○ 4
○ 6
⬭ 8

© Harcourt

1 🔖 **3.3A** Noah has 6 apples. Lorraine has 4 apples, and Greg has 2 apples. How many apples do they have altogether? Mark your answer.

⬭ 5 apples
⬭ 7 apples
▬ 12 apples
⬭ 22 apples

6+2=8
8+4=12

3 🔖 **3.11E** Helen has a container with 4 gallons of water. Which shows how many quarts are equal to 4 gallons? Mark your answer.

⬭ 12 quarts
⬭ 14 quarts
⬭ 16 quarts
▬ 18 gallons

2 🔖 **3.4C** Which division sentence is modeled by the array? Mark your answer.

[array of dots: 5 rows × 6 columns of filled circles]

⬭ 30 ÷ 15 = 2
⬭ 30 ÷ 8 = 5
⬭ 40 ÷ 5 = 8
▬ 30 ÷ 5 = 6

4 🔖 **3.14C** The table shows the cost of tickets at the Museum of Fine Arts in Houston, Texas. Which is a rule for the table? Mark your answer.

Tickets	2	4	6	8	10
Cost	$14	$28	$42	$56	$70

⬭ Multiply the number of tickets by $3
⬭ Multiply the number of tickets by $5
⬭ Add $6 to the number of tickets
▬ Multiply the number of tickets by $7

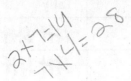

2×7=14
7×4=28

Name _____

5 ⬥ **3.4B** Which is the product?

$$\begin{array}{r} 25 \\ \times\ 3 \end{array}$$

Record your answer in the boxes below. Then fill in the bubbles. Be sure to use the correct place value.

6 ⬥ **3.13C** Gina made a tally table to record her friends' votes for their favorite sport. Which shows how many of Gina's friends voted for football as their favorite sport? Mark your answer.

Favorite Sport	
Subject	**Tally**
Baseball	ЖЖ ЖЖ II
Football	ЖЖ ЖЖ III
Soccer	ЖЖ I
Hockey	ЖЖ ЖЖ I

- ⬭ 6
- ⬭ 11
- ⬭ 12
- ⬭ 13

7 ⬥ **3.7B; 3.14C** The table shows the cost of a ticket at the Austin Museum of Art. Which shows the cost of 12 tickets? Mark your answer.

Tickets	2	4	6	8	10	12
Cost	$8	$16	$24	$32	$40	☐

- ⬭ $42
- ⬭ $44
- ⬭ $48
- ⬭ $50

8 ⬥ **3.9C; 3.14C** Which of the following letters appear to have more than one line of symmetry? Mark your answer.

- H
- ⬭ J
- ⬭ K
- ⬭ P

© Harcourt

9 🔻 **3.1A** The population of Jasper, Texas is 8,247. Which is the value of the underlined digit? Mark your answer.

⊘ 2
⊘ 20
✏ 200
⊘ 2,000

10 🔻 **3.6A; 3.16A** Jacob's train leaves at 6:00 P.M. The train stops at 6:04 P.M., then at 6:08 P.M., and then at 6:12 P.M. All 6 of the stops are the same distance apart. What time does Jacob arrive at the last stop? Mark your answer.

Stops	1	2	3	4	5	6
Times	6:04	6:08	6:12	6:16	6:20	☐

6:24

⊘ 6:14 P.M.
⊘ 6:18 P.M.
⊘ 6:20 P.M.
✏ 6:24 P.M.

11 🔻 **3.8** Melissa drew a octagon. Which could be Melissa's figure? Mark your answer.

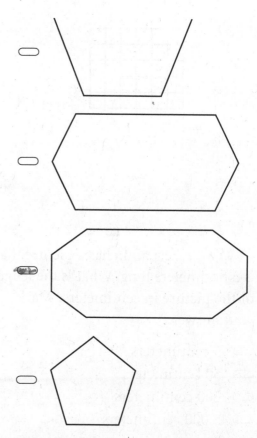

12 🔻 **3.13C** Which is the probability of pulling a white marble from the bag if all the marbles are the same size? Mark your answer.

⊘ Impossible
⊘ Unlikely
✏ Likely
⊘ Certain

13 ⭐ 3.4A Which is the value of the missing number? Mark your answer.

$$\square \times 6 = 30$$

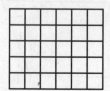

- ⬭ 1
- ⬭ 3
- ⬬ 5
- ⬭ 7

14 ⭐ 3.11A Fernando has a picture that is 5 decimeters long. What is the length of the picture in centimeters? Mark your answer.

- ⬭ 5 centimeters
- ⬬ 50 centimeters
- ⬭ 500 centimeters
- ⬭ 5,000 centimeters

50 cm

15 ⭐ 3.1C Sanjay finds 4 pennies, 1 nickel, 2 dimes, and 2 quarters. Which is the total amount Sanjay found? Mark your answer.

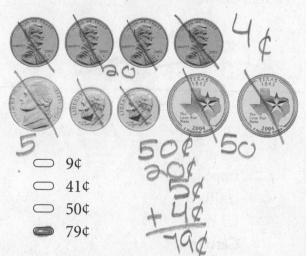

4¢

5

50¢ 50
20¢
5¢
+ 4¢
79¢

- ⬭ 9¢
- ⬭ 41¢
- ⬭ 50¢
- ⬬ 79¢

16 ⭐ 3.9C; 3.14C Which figure would complete the given letter so that the line is a line of symmetry? Mark your answer.

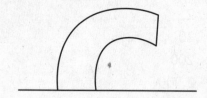

- ⬭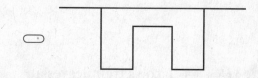
- ⬭
- ⬬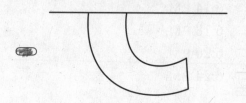
- ⬭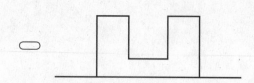

© Harcourt

17 🔹 **3.1B** Ashley's school has 321 second-grade students, 298 third-grade students, and 278 fourth-grade students. Which shows the numbers ordered from least to greatest? Mark your answer.

- ⊂⊃ 321; 298; 278
- ⊂⊃ 298; 278; 321
- ⬤ 278; 298; 321
- ⊂⊃ 321; 278; 298

278 298 398
21

18 🔹 **3.11B** Tracy has a picture. Two sides measure 10 inches and the other two sides measure 6 inches. Which is the perimeter of the picture? Mark your answer.

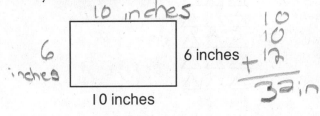

10
10
+ 12
32 in

- ⊂⊃ 6 inches
- ⊂⊃ 10 inches
- ⊂⊃ 16 inches
- ⬤ 32 inches

19 🔹 **3.5B** Use compatible numbers to estimate the difference between 243 and 124. Mark your answer.

- ⊂⊃ 200 − 100 = 100
- ⬤ 240 − 120 = 120
- ⊂⊃ 275 − 125 = 150
- ⊂⊃ 230 − 120 = 110

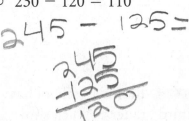

245 − 125 =
245
−125
120

20 🔹 **3.13C** Robin is going to pull one marble out of the bag. What is the probability of pulling a grey marble out of the bag if all the marbles are the same size? Mark your answer.

- ⬤ Impossible
- ⊂⊃ Unlikely
- ⊂⊃ Likely
- ⊂⊃ Certain

© Harcourt

21 3.8 How many sides does a hexagon have? Mark your answer.

○ 5
◉ 6
○ 7
○ 8

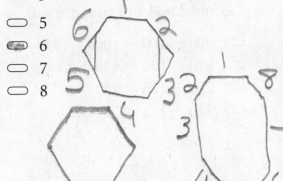

22 3.1A Rashida's school has 247 third graders. Which shows 247 written in word form? Mark your answer.

○ Twenty-four
○ Two hundred forty
◉ Two hundred forty-seven
○ Two hundred forty-seven thousand

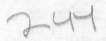

23 3.6A; 3.16A How many triangles will be in the next row? Mark your answer.

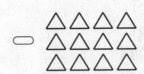

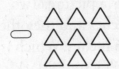

24 ⬇ **3.13B** The bar graph shows the results of Jose's survey of favorite foods. Which shows how many more people chose Burritos than Meatloaf? Mark your answer.

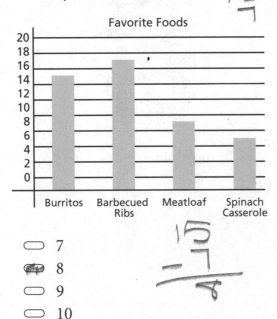

Favorite Foods

- ⬭ 7
- ⬬ 8
- ⬭ 9
- ⬭ 10

25 ⬇ **3.6A; 3.16A** Use the hundred chart. Which of the following set of numbers shows skip counting 5?

1	2	3	4	5	6	7	8	9	10
11	12	13	14	15	16	17	18	19	20
21	22	23	24	25	26	27	28	29	30
31	32	33	34	35	36	37	38	39	40
41	42	43	44	45	46	47	48	49	50
51	52	53	54	55	56	57	58	59	60
61	62	63	64	65	66	67	68	69	70
71	72	73	74	75	76	77	78	79	80
81	82	83	84	85	86	87	88	89	90
91	92	93	94	95	96	97	98	99	100

- ⬭ 7, 21, 35
- ⬭ 12, 24, 30
- ⬬ 15, 25, 35
- ⬭ 18, 27, 36

26 ⬇ **3.8; 3.14C** Which of the following can **NOT** have a bottom, side, or top view that looks like the figure? Mark your answer.

- ⬭ Cylinder
- ⬭ Rectangular prism
- ⬭ Cone
- ⬬ Sphere

27 🏴 **3.13A** Lydia wants to make a bar graph to show how many CDs she has collected. Which type of CD would have the longest bar? Mark your answer.

- ⬬ 65 rock and roll CDs
- ⬭ 45 country CDs
- ⬭ 26 hip-hop CDs
- ⬭ 32 jazz CDs

© Harcourt

28 🟤 3.3A Which addition sentence is shown by the model? Mark your answer.

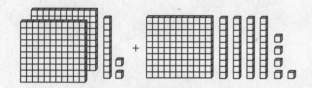

- ⬭ 223 + 154 = 377
- ⬭ 122 + 100 = 222
- ⬭ 145 + 145 = 290
- ⬬ 212 + 145 = 357

212+145

213
+145
357

29 🟤 3.13C What is the probability of pulling a black, or grey marble out of the bag if all the marbles are the same size? Mark your answer.

- ⬭ Impossible
- ⬭ Unlikely
- ⬭ Likely
- ⬬ Certain

30 3.4C Billy bought 25 pens. The pens came in packages of 5. Which shows how many packages Billy bought? Mark your answer.

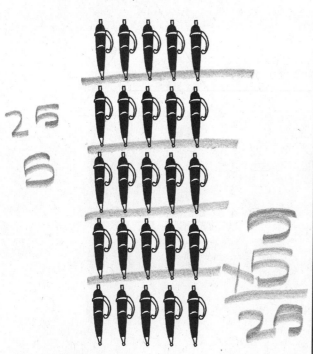

- ● 5 packages
- ○ 8 packages
- ○ 10 packages
- ○ 12 packages

31 3.6A Which are the next two numbers in the pattern?

5, 6, 8, 11, 15, 20, ☐, ☐

- ○ 20, 30
- ○ 25, 32
- ○ 26, 33
- ○ 30, 34

32 3.8 Which correctly names this triangle? Mark your answer.

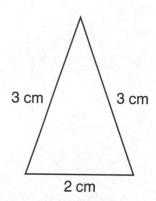

- ○ Equilateral
- ○ Right
- ○ Isosceles
- ○ Scalene

Name _____

33 3.9A Which figure appears to be congruent to the figure shown? Mark your answer.

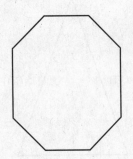

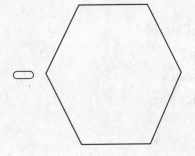

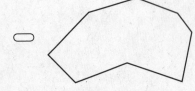

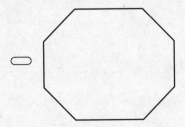

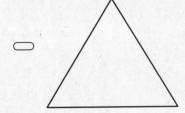

34 3.9C Which shows how many lines of symmetry the figure appears to have? Mark your answer.

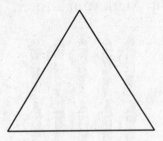

⬭ 1
⬭ 2
⬭ 3
⬭ 4

© Harcourt

35 🔹 3.11F Which is the volume of the prism? Mark your answer.

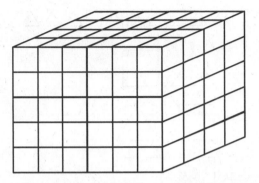

⬭ 30 cubic units
⬭ 60 cubic units
⬭ 120 cubic units
⬭ 564 cubic units

36 🔹 3.11D Mackenzie has a book that weighs 3 pounds. Which shows how many ounces are equal to 3 pounds.

⬭ 16 ounces
⬭ 32 ounces
⬭ 36 ounces
⬭ 48 ounces

37 🔹 3.4A Which multiplication sentence does the array show? Mark your answer.

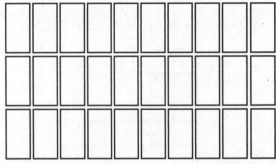

⬭ 3 + 10 = 13
⬭ 3 × 3 = 9
⬭ 3 × 10 = 30
⬭ 10 × 10 = 100

38 🔹 3.11E; 3.14A Ryan's mom is giving Ryan liquid medicine in a spoon. Which shows about how much medicine the spoon can hold? Mark your answer.

⬭ about 1 mL
⬭ about 100 mL
⬭ about 1 L
⬭ about 100 L

© Harcourt

39 🔸 **3.11C** Gino arranges 1-inch square tiles to form a letter L. Which is the area of the L? Mark your answer.

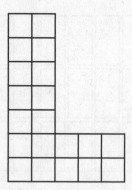

- ⬭ 12 square inches
- ⬭ 16 square inches
- ⬭ 18 square inches
- ⬭ 20 square inches

40 🔸 **3.6A** Which shows the next two figures in the pattern?

△□△△□△△△□△△ ___

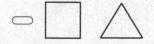

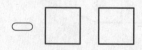

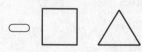

Name _____

STANDARDS PRACTICE

3.1A use place value to read, write, and describe the value of whole numbers through 9,999

1 Gina's elementary school in Fort Worth has 864 students. Which shows 864 written in expanded form? Mark your answer.

  ⬭ 8 + 6 + 4
  ⬭ 800 + 60 + 4
  ⬭ 800 + 6 + 4
  ⬭ 86 + 4

2 Which is the value of the underlined digit in 69,281? Mark your answer.

  ⬭ 9
  ⬭ 900
  ⬭ 9,000
  ⬭ 90,000

3 Seth's school has 547 third-grade students. Which shows 547 written in word form? Mark your answer.

  ⬭ five hundred forty-seven hundred
  ⬭ five hundred forty-seven
  ⬭ forty-seven
  ⬭ five thousand forty-seven

4 Guadalupe Peak is the highest point in Texas and is 8,749 feet above sea level. What is the value of the underlined digit?

  _____

5 The driving distance from San Antonio, Texas to San Francisco, California is about 1,734 miles. Write 1,734 written in expanded form.

  _____

6 Texas covers a total area of 268,581 square miles. What is the value of the underlined digit?

  _____

© Harcourt

7 The greatest straight-line distance in a north-south direction through Texas is 801 miles. Which is the value of the 1 in 801? Mark your answer.

- ⬭ 1
- ⬭ 10
- ⬭ 100
- ⬭ 1,000

8 In 2000, the estimated population of Wichita Falls, Texas was 104,197. Which shows 104,197 written in word form? Mark your answer.

- ⬭ one hundred four thousand
- ⬭ one hundred four thousand, one hundred ninety-seven
- ⬭ one hundred four thousand, one hundred ninety-seven hundred
- ⬭ one hundred four hundred, one hundred ninety-seven

9 Nathan's elementary school in Austin has 1,325 students. What is 1,325 written in expanded form?

_____

10 Greg's store in Lubbock had 2,8<u>9</u>5 visitors during the first week in December. What is the value of the underlined digit?

_____

11 Henry drove 689 miles during a trip visiting different Texas towns. What is the value of the 6 in 689?

_____

© Harcourt

⭐ **3.1B**   use place value to compare and order whole numbers through 9,999

1 In 2000, the population of Bangs, Texas was 1,620, the population of Barton Creek, Texas was 1,589, and the population of Beach City, Texas was 1,645. Which shows the numbers ordered from greatest to least? Mark your answer.

    ⬭ 1,589; 1,620; 1,645
    ⬭ 1,645; 1,620; 1,589
    ⬭ 1,645; 1,589; 1,620
    ⬭ 1,589; 1,645; 1,620

2 Rachael is reading a book with 215 pages. Frank is reading a book with 209 pages. Which statement is correct? Mark your answer.

    ⬭ 209 = 215
    ⬭ 209 > 215
    ⬭ 215 = 209
    ⬭ 215 > 209

3 Which number is less than point *G* on the number line? Mark your answer.

    ⬭ 9
    ⬭ 11
    ⬭ 12
    ⬭ 14

4 Janette's school in Flower Mound, Texas, has 343 first graders, 356 second graders, and 339 third graders. Order these numbers from least to greatest.

_____

5 Lewis has a collection of 137 trading cards. Michael has a collection of 151 trading cards. Write >, <, or = to make the statement true.

137 ◯ 151

6 Barbara's store had 3,751 visitors in March, 3,915 visitors in April, and 3,802 visitors in May. Order these numbers from greatest to least.

_____

© Harcourt

7 Which number is greater than point *F* on the number line? Mark your answer.

⬭ 11

⬭ 8

⬭ 3

⬭ 1

8 Hector throws a football 93 feet on his first toss. He throws the football 92 feet on his second toss, and 101 feet on his third toss. Which shows the numbers ordered from greatest to least? Mark your answer.

⬭ 92, 93, 92

⬭ 101, 93, 92

⬭ 93, 101, 92

⬭ 101, 92, 93

9 Phoebe scored 128 points playing a board game. Jason scored 119 points. Which statement is correct? Mark your answer.

⬭ 119 > 128

⬭ 128 < 119

⬭ 128 > 119

⬭ 119 = 128

10 Eric's school has 294 fourth-grade students, 307 fifth-grade students, and 276 first-grade students. Order the numbers from greatest to least.

_____

11 What is a number that is less than point *Q* but greater than 50 on the number line?

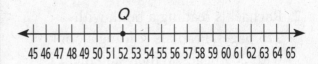

_____

12 George took 168 pictures with his camera on his vacation. Beth took 168 pictures with her camera on her vacation. Write <, >, or = to make the statement true.

3.1C determine the value of a collection of coins and bills

1 What is the amount shown? Mark your answer.

- ⬭ $7.75
- ⬭ $8.15
- ⬭ $12.15
- ⬭ $13.75

2 Alex finds 1 penny, 3 nickels, 2 dimes, and 2 quarters in her pocket. What is the total amount of money Alex found? Mark your answer.

- ⬭ 8¢
- ⬭ 86¢
- ⬭ 91¢
- ⬭ 96¢

3 Ray bought a package of Houston Astros baseball cards. He paid with a ten-dollar bill and received 2 one-dollar bills, 3 quarters, and 1 dime in change. How much did the baseball cards cost?

_____

4 Teresa and Hector each bought a type of sandwich for lunch. They paid the amounts shown below. Write <, >, or = to make the statement true.

$4.85 ◯ $4.80

© Harcourt

5 Which is the missing number? Mark your answer.

2 quarters + ☐ nickels = $1.00

- ⬭ 5
- ⬭ 10
- ⬭ 15
- ⬭ 20

6 Gina has only quarters. She has less than $1.00. Which amount could Gina have? Mark your answer.

- ⬭ $0.45
- ⬭ $0.50
- ⬭ $0.95
- ⬭ $1.25

7 Jamie bought a t-shirt on sale. He paid with a ten-dollar bill and received a one-dollar bill, 3 dimes, and 7 pennies in change. How much did the t-shirt cost?

8 What is the amount?

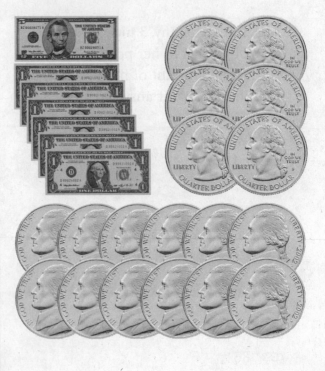

3.2A construct concrete models of fractions

1 Walter is making a tile border. What fraction of the figure is shaded? Mark your answer.

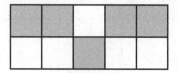

- ⬭ three fifths
- ⬭ three tenths
- ⬭ five tenths
- ⬭ seven tenths

2 Joseph is playing a game using the spinner shown below. What fraction of the spinner is shaded? Mark your answer.

- ⬭ one eighth
- ⬭ four eighths
- ⬭ six eighths
- ⬭ four fourths

3 Michelle is playing a game using the board shown below. What fraction of the board is shaded?

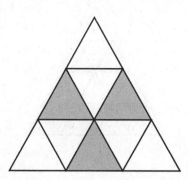

_____

4 Adam brings a pie to a dinner. The pie is cut into equal sections. The shaded portion shows the number of slices of pie that were eaten. What fraction of the pie is left over?

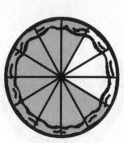

_____

5 Brian is playing a game using the board shown below. What fraction of the board is shaded? Mark your answer.

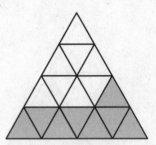

○ $\frac{2}{16}$

○ $\frac{6}{16}$

○ $\frac{8}{16}$

○ $\frac{9}{16}$

6 Lucy's family has pizza for dinner. The pizza is cut into equal sections. The shaded portion shows the number of slices of pizza that were eaten. What fraction of the pizza is left over? Mark your answer.

○ one eighth
○ one seventh
○ seven eighths
○ eight sevenths

7 David is making a tile border. What fraction of the figure is shaded?

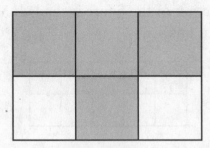

_____

8 Georgia is playing a game using the counters. What fraction of the counters are shaded?

_____

© Harcourt

🔷 **3.2B** compare fractional parts of whole objects or sets of objects in a problem situation using concrete models

1 Heather and Davis order a pumpkin pie. The pie is cut into 10 equal pieces. Davis eats $\frac{3}{10}$ of the pieces. Heather eats a smaller amount than Davis. Which fraction shows how much of the pie Heather ate? Mark your answer.

- ⬭ $\frac{2}{10}$
- ⬭ $\frac{3}{10}$
- ⬭ $\frac{4}{10}$
- ⬭ $\frac{7}{10}$

2 Joseph and Franklin order a pizza. The pizza is cut into 8 equal pieces. Joseph eats $\frac{1}{2}$ of the pieces. Franklin eats an equal amount. Which fraction shows how much of the pizza Franklin ate? Mark your answer.

- ⬭ $\frac{1}{8}$
- ⬭ $\frac{3}{4}$
- ⬭ $\frac{4}{8}$
- ⬭ $\frac{3}{8}$

3 Phil has a package of 12 trading cards. He uses $\frac{1}{3}$ of the cards in a game. He gives away a smaller amount of the cards to his friend. What is a fraction that shows how many cards Phil gave away.

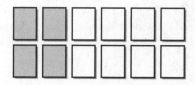

4 Judith brought a cake to a birthday party. The cake was cut into 16 equal pieces. Judith eats $\frac{1}{8}$ of the pieces. Mary eats a smaller amount than Judith. What is a fraction that shows how much of the cake Mary ate.

5 Cary has a package of 10 cookies. She eats $\frac{2}{5}$ of the cookies. She gives away a larger amount of the cookies to her friends. Which fraction shows how many cookies Cary gave away? Mark your answer.

○ $\frac{4}{10}$

○ $\frac{1}{4}$

○ $\frac{1}{2}$

○ $\frac{1}{5}$

6 Melissa bought a package of strawberries. There are 15 strawberries in the package. She eats $\frac{2}{5}$ of the strawberries. She uses a larger amount to decorate the top of a cake. Which fraction shows how many strawberries Melissa used to decorate the cake? Mark your answer.

○ $\frac{1}{15}$

○ $\frac{9}{15}$

○ $\frac{6}{15}$

○ $\frac{2}{5}$

7 Jessica has a package of 9 marbles. Of the marbles, $\frac{1}{3}$ are gray. A smaller amount of the marbles are black. Write a fraction that could show how many marbles are black.

_____

8 Harry ordered a pizza. The pizza was cut into 8 equal pieces. Harry wanted $\frac{1}{4}$ of the pizza covered with sausage and a larger amount covered with pepperoni. Write a fraction that could show how much of the pizza was covered with pepperoni.

_____

© Harcourt

3.2C use fraction names and symbols to describe fractional parts of whole objects or sets of objects

1 Beth has 12 marbles. Of the marbles, $\frac{3}{12}$ are white. Which model shows what fraction of the marbles are white? Mark your answer.

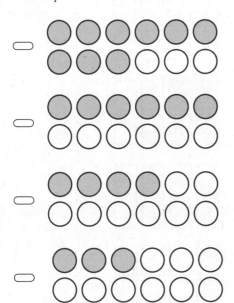

3 Jake has 10 trading cards. Of the cards, $\frac{1}{2}$ are turned over so that the shaded side is up. Which model shows what fraction of the cards are turned over? Mark your answer.

2 Lars orders a pizza. The pizza has 8 equal slices. Of the slices, $\frac{1}{4}$ are covered with sausage. How many slices are covered in sausage?

_____

Name _____

STANDARDS PRACTICE

4 Tracy has 6 tennis balls. Of the tennis balls, $\frac{2}{6}$ are white. Which model shows what fraction of the tennis balls is white? Mark your answer.

○

○

○

○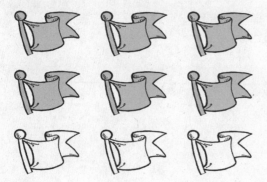

6 Horace has 8 game cards. Of the cards, $\frac{3}{4}$ are turned over so that the shaded side is up. Which model shows what fraction of the cards are turned over? Mark your answer.

○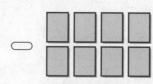

○

○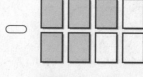

○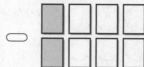

5 Beth has 9 flags. Of the flags, $\frac{2}{3}$ are shaded. How many flags are shaded?

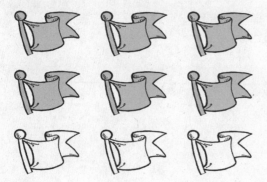

Practice for TAKS Success   58   **Standards Practice**

3.2D construct concrete models of equivalent fractions for fractional parts of whole objects

1 Uli has a set of 12 trading cards. Of the cards, 6 are turned over. Which shows what fraction of the cards are turned over? Mark your answer.

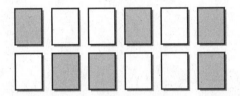

○ $\frac{5}{12}$

○ $\frac{1}{2}$

○ $\frac{1}{6}$

○ $\frac{8}{12}$

2 Victor is playing a game with a spinner. What fraction is equal to the part of the spinner that is shaded? Mark your answer.

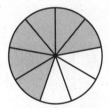

○ $\frac{1}{4}$

○ $\frac{1}{3}$

○ $\frac{1}{2}$

○ $\frac{2}{3}$

3 Matt has 8 coins and 4 of the coins are quarters. The rest are pennies. What fraction of the coins are quarters?

_____

4 Jason has a pizza. The shaded portion is the part that has extra cheese. What fraction is equal to the part of the pizza that is shaded?

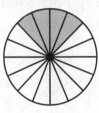

_____

© Harcourt

5 Karen has a set of 10 game cards, 4 of which are turned over. Which shows what fraction of the cards are turned over written in simplest form? Mark your answer.

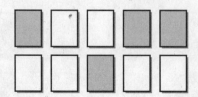

○ $\frac{2}{5}$

○ $\frac{1}{4}$

○ $\frac{1}{2}$

○ $\frac{6}{10}$

6 Jeffrey is playing a game with a spinner. What fraction is equal to the part of the spinner that is shaded? Mark your answer.

○ $\frac{1}{4}$

○ $\frac{1}{3}$

○ $\frac{1}{2}$

○ $\frac{2}{3}$

7 Helen has 15 pennies and 10 of the pennies are heads up. The rest of the pennies are tails up. What fraction of the pennies are heads up?

_____

8 Michael has a pizza. The shaded portion shows the part of the pizza that has pepperoni. What fraction is equal to the part of the pizza that is shaded?

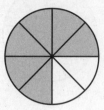

_____

© Harcourt

⬥ **3.3A**   **model addition and subtraction using pictures, words, and numbers**

1 Julia bought 7 songs for her music player. Marsha bought 9 songs for her music player. Jay bought 4 songs for his music player. How many total songs did they buy? Mark your answer.

   ⬭ 11

   ⬭ 13

   ⬭ 16

   ⬭ 20

3 What number makes this number sentence true?

$$\square + 8 = 13$$

_____

4 What number makes this number sentence true?

$$11 - 5 = \square$$

_____

2 Which addition sentence is shown by the model? Mark your answer.

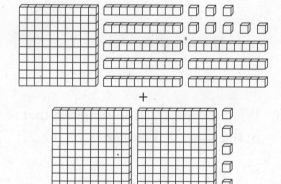

   ⬭ 180 + 200 = 380

   ⬭ 188 + 200 = 388

   ⬭ 188 + 205 = 393

   ⬭ 200 + 200 = 400

5 What number makes this number sentence true?

$$11 + \square = 18$$

_____

6 Which subtraction sentence is shown by the model? Mark your answer.

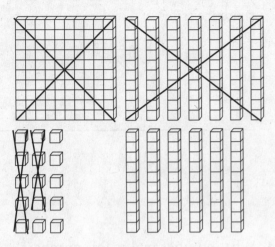

- ⬭ $245 - 179 = 66$
- ⬭ $245 - 180 = 65$
- ⬭ $250 - 180 = 70$
- ⬭ $250 - 179 = 69$

7 Which number sentence is in the same fact family as 63, 49, 14? Mark your answer.

- ⬭ $14 + 63 = 77$
- ⬭ $49 - 14 = 35$
- ⬭ $63 + 14 = 77$
- ⬭ $49 + 14 = 63$

8 What number makes this number sentence true?

$$\square - 5 = 3$$

_____

9 Jack, Ben, and Mila were playing a game with picture cards.
Jack drew a card with 8 circles on it.
Ben drew a card with 5 circles on it.
Mila drew a card with 9 circles on it.
How many total circles did all three cards have?

_____

10 What is a subtraction sentence that is in the same fact family as 18, 40, 22?

_____

3.3B select addition or subtraction and use the operation to solve problems involving whole numbers through 999

1 Mrs. Rovitti's science class collected 28 insects for a project on Monday, 37 insects on Tuesday, and 41 insects on Wednesday. Which shows how many total insects the class collected? Mark your answer.

◯ 65
◯ 78
◯ 87
◯ 106

2 Heather has 59 marbles. A friend gives her 68 more marbles. How many total marbles does Heather have now? Mark your answer.

◯ 9
◯ 126
◯ 127
◯ 172

3 In 2000, the population of Alice Acres, Texas was 491 and the population of Annona, Texas was 282. What was the total population of both towns? Mark your answer.

◯ 209
◯ 773
◯ 783
◯ 873

4 Adrian took a survey of 800 residents in Dallas, Texas and asked them to list their favorite sport. Of the people surveyed, 376 chose football. How many of the people surveyed chose a different sport other than football?

_____

5 George has 68 baseball cards in his collection. Richard gives George 23 more baseball cards. How many total baseball cards does George have now?

_____

6 Laura read 43 pages in her book on Monday, 38 pages on Tuesday, and 47 pages on Wednesday. How many total pages did Laura read during the 3 days? Mark your answer.

_____

© Harcourt

7 In 2000, the population of Celeste, Texas was 817 and the population of Bryson was 528. How many more people lived in Celeste in 2000? Mark your answer.

⊂⊃ 189
⊂⊃ 282
⊂⊃ 289
⊂⊃ 299

8 Brianna has 57 beads. She has 39 round beads. The rest of the beads are square. How many of the beads are square? Mark your answer.

⊂⊃ 8
⊂⊃ 18
⊂⊃ 19
⊂⊃ 23

9 Fredrick took a survey of 600 residents in Garland, Texas and asked them to list their favorite sport. Of the people, 263 surveyed chose football. How many of the people surveyed chose a different sport other than football? Mark your answer.

⊂⊃ 337
⊂⊃ 357
⊂⊃ 437
⊂⊃ 447

10 A museum in Austin, Texas had 765 visitors on Tuesday. On Wednesday, the museum had 597 visitors. How many more visitors did the museum have on Tuesday?

_____

11 Kat bicycled for 43 minutes on Thursday, 51 minutes on Friday, and 54 minutes on Saturday. How many total minutes did Kat bicycle during the 3 days?

_____

12 Jerry has 17 comic books. A friend gives him 19 more comic books. How many comic books does Jerry have now?

_____

© Harcourt

◆ 3.4A **learn and apply multiplication facts through 12 by 12 using concrete models and objects**

1 Which multiplication sentence is modeled? Mark your answer.

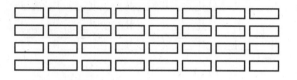

  ⬭ 4 + 8 = 12
  ⬭ 4 × 4 = 16
  ⬭ 4 × 8 = 32
  ⬭ 8 × 8 = 64

3 Gigi is making a design with 3 hexagons. Each hexagon has 6 sides. Write the multiplication sentence that shows how many sides 3 hexagons have.

_____

2 Which multiplication sentence is modeled by the array? Mark your answer.

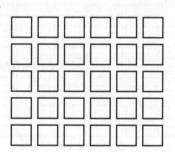

  ⬭ 5 + 6 = 11
  ⬭ 5 × 6 = 30
  ⬭ 5 × 5 = 50
  ⬭ 5 × 6 = 56

4 What multiplication sentence is modeled?

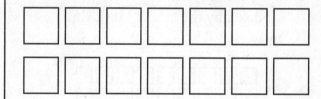

_____

© Harcourt

Name _____

5 Jenna is making a design with 3 squares. Each square has 4 sides. Which multiplication sentence shows how many sides 3 squares have? Mark your answer.

- ⬭ 4 × 3 = 12
- ⬭ 3 × 3 = 9
- ⬭ 4 × 4 = 16
- ⬭ 12 × 3 = 36

7 What multiplication sentence is modeled by the array?

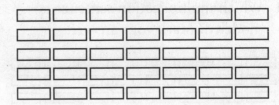

_____

6 Which multiplication sentence is modeled by the array? Mark your answer.

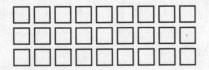

- ⬭ 9 + 3 = 12
- ⬭ 9 × 3 = 27
- ⬭ 9 + 9 = 18
- ⬭ 9 × 9 = 81

8 What multiplication sentence is modeled by the array?

_____

© Harcourt

3.4B **solve and record multiplication problems**

1 Chris used 9 cans of stewed tomatoes to make his famous pasta sauce. Each can had 5 tomatoes. How many tomatoes did Chris put into his sauce? Mark your answer.

- ○ 14
- ○ 27
- ○ 36
- ○ 45

2 John, Mick and James go to a pizza parlor. They each pay $2 for a slice of pizza. Which shows the total amount they paid? Mark your answer.

- ○ $2
- ○ $4
- ○ $6
- ○ $10

3 What is the value of the missing number? Mark your answer.

$$\square \times 1 = 8$$

- ○ 8
- ○ 9
- ○ 10
- ○ 18

4 The high school marching band has 9 rows and 12 musicians in each row. How many total musicians are in the marching band?

_____

5 What is the value of the missing number?

$$6 \times \square = 42$$

_____

6 Jalissa colored 3 pictures each day for 7 days. How many total pictures did Jalissa color?

_____

7 Tina rides her favorite rollercoaster 6 times. Each time the ride costs $3. Which multiplication sentence shows how to find the total fees for 6 rides? Mark your answer.

- ⬭ $3 × 3 = $9
- ⬭ $3 × 6 = $18
- ⬭ $3 + 6 = $9
- ⬭ $6 × 6 = $36

8 Jasmine jogs for 3 miles each day for 8 days. How many total miles did Jasmine jog? Mark your answer.

- ⬭ 11
- ⬭ 18
- ⬭ 24
- ⬭ 32

9 Which is the value of the missing number? Mark your answer.

$$\square × 5 = 60$$

- ⬭ 12
- ⬭ 15
- ⬭ 20
- ⬭ 25

10 What is the missing number?

$$\square × 10 = 90$$

_____

11 Becky cleaned her room 5 times last month. She spent 2 hours cleaning it each time. How many total hours did Becky clean her room last month?

_____

12 What multiplication sentence is related to the addition sentence?

$$3 + 3 + 3 + 3$$

_____

© Harcourt

13 A group of 5 students each borrow 2 books from the library. How many total books did they borrow? Mark your answer.

- ⬯ 5
- ⬯ 7
- ⬯ 10
- ⬯ 52

14 What is the missing factor? Mark your answer.

$$10 \times \square = 0$$

- ⬯ 0
- ⬯ 1
- ⬯ 10
- ⬯ 12

15 Joseph has 6 sets of colored pencils. Each set has 9 different colored pencils. How many total pencils does Joseph have? Mark your answer.

- ⬯ 3
- ⬯ 15
- ⬯ 45
- ⬯ 54

16 Helen makes 7 photocopies at the library. Each photocopy costs 12¢. How much did Helen pay for all of the copies?

_____

17 What is the missing factor?

$$3 \times \square = 33$$

_____

18 Casey has 2 packages of bread. Each package has 16 slices. How many slices of bread does Casey have?

_____

Name _____

19 A group of 4 students go to a water park. Each ticket cost $7. Which shows how much the students spent? Mark your answer.

- $12
- $20
- $28
- $32

20 The Hardware Store has a display of boards. The boards are in 12 rows and 9 boards in each row. How many total boards does the display have? Mark your answer.

- 108
- 110
- 111
- 208

21 Betty's bakery buys 9 bags of 8 apples each day to make apple pies. Which shows how many total apples the bakery buys each day? Mark your answer.

- 17
- 36
- 54
- 72

22 Helen bicycles 5 miles each day for one week. How many miles did Helen bike in her first week?

_____

23 Julia and her 6 family members go out for ice cream. Each ice cream costs $2. What was the total amount the family spent?

_____

24 What is the missing factor?

$$3 \times \square = 30$$

_____

© Harcourt

3.4C use models to solve division problems and use number sentences to record the solutions

1 A group of 108 students visit Johnson Space Center in Houston, Texas. The students tour the center in groups of 9. Which shows how many groups of students toured the center? Mark your answer.

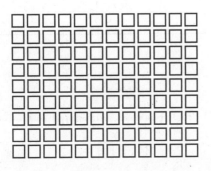

⊂⊃ 8   ⊂⊃ 12

⊂⊃ 11   ⊂⊃ 75

2 A group of 48 students are working in groups on a project at school. The students split into equal groups of 8 students. How many groups will they have? Mark your answer.

```
○○○○○○
○○○○○○
○○○○○○
○○○○○○
○○○○○○
○○○○○○
○○○○○○
○○○○○○
```

⊂⊃ 4   ⊂⊃ 7

⊂⊃ 6   ⊂⊃ 9

3 Christy bought 40 stickers. The stickers came in packages of 5. How many packages were there?

_____

4 Larry jogs 16 miles in one week to start training for the Houston Marathon. He jogged 4 miles per day. How many total days did Larry jog during the first week?

= = = =
= = = =

_____

5 Which is the missing number? Mark your answer.

$$\square \div 3 = 6$$

⬭ 6   ⬭ 2
⬭ 9   ⬭ 18

6 Danielle wants to divide a group of pencils into 5 equal groups. She opens her box and finds 0 pencils. How many pencils will be in each group? Mark your answer.

⬭ 0   ⬭ 5
⬭ 1   ⬭ 10

7 A group of students paid a total of $72 to buy tickets for a movie. Each ticket cost $8. Which shows how many students were in the group? Mark your answer.

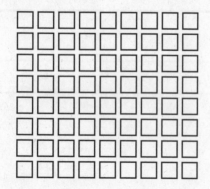

⬭ 6   ⬭ 9
⬭ 8   ⬭ 12

8 What division sentence is related to the given set of subtraction sentences?

$$48 - 12 = 36$$
$$36 - 12 = 24$$
$$24 - 12 = 12$$
$$12 - 12 = 0$$

_____

9 What is the missing number?

$$84 \div \square = 12$$

_____

10 Jane has 54 photographs in a photo album. The album fits 6 photos per page. How many pages in the album are filled?

_____

© Harcourt

3.5A round whole numbers to the nearest ten or hundred to approximate reasonable results in problem situations

1 Tony's mother is reading a book that has 178 pages. Which shows the number of pages rounded to the nearest hundred? Mark your answer.

- ⬭ 100
- ⬭ 170
- ⬭ 180
- ⬭ 200

2 In the year 2000, Austin Elementary School had 528 students. Which shows the number of students rounded to the nearest ten? Mark your answer.

- ⬭ 500
- ⬭ 520
- ⬭ 530
- ⬭ 600

3 The driving distance from Austin to Corpus Christi is about 202 miles. Which shows the distance rounded to the nearest hundred? Mark your answer.

- ⬭ 200
- ⬭ 210
- ⬭ 300
- ⬭ 302

4 Write the number represented by point X rounded to the nearest ten.

_____

5 Mr. Gordon's 3rd grade class has 33 students. Write the number of students rounded to the nearest ten.

_____

6 The driving distance from Corpus Christi to Perryton is about 693 miles. Write the distance rounded to the nearest hundred.

_____

© Harcourt

7   Bobby is reading a book that has
    247 pages. Which shows the number
    of pages rounded to the nearest ten?
    Mark your answer.

    ○ 200

    ○ 240

    ○ 250

    ○ 300

8   Which shows the number represented
    by point *Y* rounded to the nearest ten?
    Mark your answer.

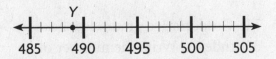

    ○ 400

    ○ 480

    ○ 490

    ○ 500

9   Abernathy, Texas has a population of
    about 2,839. Which shows the number
    of people rounded to the nearest
    hundred? Mark your answer.

    ○ 2,000

    ○ 2,800

    ○ 2,900

    ○ 3,000

10 Write the number represented by
    point *B* rounded to the nearest ten.

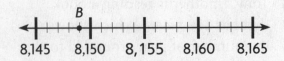

    _____

11 Bayview, Texas has a population of 323.
    Write the number of people rounded
    to the nearest ten.

    _____

12 The driving distance from Austin
    to El Paso is about 576 miles. Write
    the distance rounded to the nearest
    hundred.

    _____

© Harcourt

3.5B use strategies including rounding and compatible numbers to estimate solutions to addition and subtraction problems

1 El Paso is about 209 miles from Pecos. Pecos is about 96 miles from Midland. Round to the nearest hundred the total distance from El Paso to Pecos and then from Pecos to Midland. Mark your answer.

   ⬭ about 100 miles
   ⬭ about 200 miles
   ⬭ about 300 miles
   ⬭ about 400 miles

2 Use compatible numbers to estimate the sum of 278 and 202? Mark your answer.

   ⬭ 200 + 200 = 400
   ⬭ 280 + 200 = 480
   ⬭ 280 + 210 = 490
   ⬭ 400 + 200 = 600

3 Houston is about 268 miles from Fort Worth. Fort Worth is about 107 miles from Cisco. Round to the nearest hundred the total distance from Houston to Fort Worth and then from Fort Worth to Cisco. Mark your answer.

   ⬭ 200
   ⬭ 300
   ⬭ 370
   ⬭ 400

4 Use compatible numbers to estimate the sum of 218 and 551.

   _____

5 The trip from Odessa to Beaumont is 663 miles. The trip from Odessa to Houston is 578 miles. Round to the nearest hundred how much longer the trip to Beaumont is.

   _____

6 Use compatible numbers to estimate the difference between 731 and 277.

   _____

© Harcourt

7 The trip from Austin to Dallas is 196 miles. The trip from Austin to Houston is 162 miles. Round to the nearest ten how much longer the trip to Houston is. Mark your answer.

- ⬭ about 40 miles
- ⬭ about 50 miles
- ⬭ about 100 miles
- ⬭ about 140 miles

8 Use compatible numbers to estimate the difference between 427 and 223. Mark your answer.

- ⬭ $400 - 200 = 200$
- ⬭ $425 - 100 = 325$
- ⬭ $425 - 225 = 200$
- ⬭ $450 - 225 = 225$

9 Use compatible numbers to estimate the sum of 678 and 148? Mark your answer.

- ⬭ $600 + 100 = 700$
- ⬭ $625 + 100 = 725$
- ⬭ $675 + 150 = 825$
- ⬭ $700 + 200 = 900$

10 Use compatible numbers to estimate the sum of 373 and 326.

_____

11 Bells, Texas has a population of 1,190. Buffalo, Texas has a population of 1,834. Round to the nearest hundred to estimate how many more people are in Buffalo, Texas.

_____

12 Use compatible numbers to estimate the difference between 981 and 719.

_____

© Harcourt

Name _____

3.6A identify and extend whole-number and geometric patterns to make predictions and solve problems

1 Which is the next number in the pattern? Mark your answer.

66, 58, 50, 42, ☐

- ◯ 28
- ◯ 30
- ◯ 32
- ◯ 34

2 Which are the next two numbers in the pattern?

2, 4, 6, 2, 4, 6, 2, ☐, ☐

- ◯ 2, 4
- ◯ 6, 2
- ◯ 4, 6
- ◯ 4, 2

3 Which is the next number in the pattern? Mark your answer.

56, 63, 70, 77, 84, ☐

- ◯ 90
- ◯ 91
- ◯ 94
- ◯ 97

4 Carrie drew a pattern. The pattern unit is square, circle, square, triangle. What will be the 11th and 12th figures?

_____

5 Mary's train leaves at 7:00 A.M. The train stops at 7:04 A.M., then at 7:08 A.M., and then at 7:12 A.M. All 9 of the stops are the same distance apart. What time does Mary arrive at the last stop?

Stops	1	2	3	4	5	6	7	8	9
Time	7:04	7:08	7:12						☐

_____

© Harcourt

6 Which are the next two numbers in the pattern?

3, 3, 7, 4, 3, 3, 7, 4, 3, 3, ☐, ☐

⬯ 7, 4
⬯ 3, 3
⬯ 4, 3
⬯ 3, 7

7 Greg drew a pattern. The pattern unit is hexagon, pentagon, circle, hexagon. What will the 9th and 10th figures be? Mark your answer.

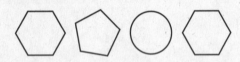

⬯ Circle, hexagon
⬯ Hexagon, pentagon
⬯ Hexagon, hexagon
⬯ Pentagon, circle

8 Which is the next number in the pattern? Mark your answer.

18, 24, 30, 36, ☐

⬯ 12
⬯ 38
⬯ 40
⬯ 42

9 What are the next two figures in the pattern? Write the shapes.

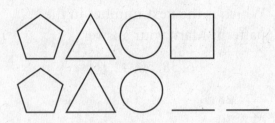

_____

10 Which shows the missing figure? Write the letter.

A

B

C

D

_____

© Harcourt

3.6B identify patterns in multiplication facts using concrete objects, pictorial models, or technology

1 Use the hundred chart. Which of the following shows multiples of 7?

1	2	3	4	5	6	7	8	9	10
11	12	13	14	15	16	17	18	19	20
21	22	23	24	25	26	27	28	29	30
31	32	33	34	35	36	37	38	39	40
41	42	43	44	45	46	47	48	49	50
51	52	53	54	55	56	57	58	59	60
61	62	63	64	65	66	67	68	69	70
71	72	73	74	75	76	77	78	79	80
81	82	83	84	85	86	87	88	89	90
91	92	93	94	95	96	97	98	99	100

⬭ 16, 28, 34
⬭ 4, 8, 12
⬭ 16, 26, 36
⬭ 28, 49, 56

2 The table shows the cost of movie tickets at the local movie theater in Bobby's town. Which of the following is a rule for the table? Mark your answer.

Tickets	3	5	7	9	11
Cost	$33	$55	$77	$99	$121

⬭ Multiply the number of tickets by $11
⬭ Multiply the number of tickets by $10
⬭ Add $30 to the number of tickets
⬭ Add $70 to the number of tickets

3 What is shown by the shaded numbers on the hundred chart?

1	2	3	4	5	6	7	8	9	10
11	12	13	14	15	16	17	18	19	20
21	22	23	24	25	26	27	28	29	30
31	32	33	34	35	36	37	38	39	40
41	42	43	44	45	46	47	48	49	50
51	52	53	54	55	56	57	58	59	60
61	62	63	64	65	66	67	68	69	70
71	72	73	74	75	76	77	78	79	80
81	82	83	84	85	86	87	88	89	90
91	92	93	94	95	96	97	98	99	100

multiples of 2 and _____

4 The table shows the cost of popcorn at a local theater. What is the cost of 8 boxes of popcorn?

Boxes	2	3	4	6	7	8
Cost	$8	$12	$16	$24	$28	☐

_____

© Harcourt

5 The table shows the cost of comic books at Rebecca's local comic shop. Which shows the cost of 9 comics? Mark your answer.

Boxes	2	4	5	7	8	9
Cost	$6	$12	$15	$21	$24	☐

- ⬭ $27
- ⬭ $28
- ⬭ $29
- ⬭ $31

6 Use the hundred chart. Which set of numbers shows multiples of 9?

1	2	3	4	5	6	7	8	9	10
11	12	13	14	15	16	17	18	19	20
21	22	23	24	25	26	27	28	29	30
31	32	33	34	35	36	37	38	39	40
41	42	43	44	45	46	47	48	49	50
51	52	53	54	55	56	57	58	59	60
61	62	63	64	65	66	67	68	69	70
71	72	73	74	75	76	77	78	79	80
81	82	83	84	85	86	87	88	89	90
91	92	93	94	95	96	97	98	99	100

- ⬭ 18, 24, 36
- ⬭ 54, 72, 96
- ⬭ 27, 36, 45
- ⬭ 9, 18, 28

7 What is shown by the shaded numbers on the hundred chart?

1	2	3	4	5	6	7	8	9	10
11	12	13	14	15	16	17	18	19	20
21	22	23	24	25	26	27	28	29	30
31	32	33	34	35	36	37	38	39	40
41	42	43	44	45	46	47	48	49	50
51	52	53	54	55	56	57	58	59	60
61	62	63	64	65	66	67	68	69	70
71	72	73	74	75	76	77	78	79	80
81	82	83	84	85	86	87	88	89	90
91	92	93	94	95	96	97	98	99	100

multiples of _____

8 The table shows the cost of boxes of stickers. What is the rule for the table?

Boxes	1	3	5	7	9
Cost	$12	$36	$60	$84	$108

_____

_____

© Harcourt

3.6C identify patterns in related multiplication and division sentences such as
$2 \times 3 = 6, 3 \times 2 = 6, 6 \div 2 = 3, 6 \div 3 = 2$

1 Which number sentence is NOT included in the same fact family as $4 \times 6 = 24$? Mark your answer.

- ⬭ $6 \times 4 = 24$
- ⬭ $96 \div 4 = 24$
- ⬭ $24 \div 4 = 6$
- ⬭ $24 \div 6 = 4$

2 Which number sentence is NOT included in the same fact family as $5 \times 9 = 45$? Mark your answer.

- ⬭ $9 \times 5 = 45$
- ⬭ $45 \div 9 = 5$
- ⬭ $45 \times 5 = 225$
- ⬭ $45 \div 5 = 9$

3 Which number sentence is NOT included in the same fact family as $35 \div 7 = 5$? Mark your answer.

- ⬭ $35 \div 5 = 7$
- ⬭ $5 \times 7 = 35$
- ⬭ $7 \times 5 = 35$
- ⬭ $35 \times 2 = 70$

4 What is a multiplication sentence in the same fact family as $24 \div 6 = 4$?

_____

5 What is a division sentence in the same fact family as $8 \times 9 = 72$?

_____

6 What is a division sentence in the same fact family as $11 \times 4 = 44$?

_____

7 Which division sentence is related to $11 \times 4 = 44$? Mark your answer.

- ⬭ $88 \div 2 = 44$
- ⬭ $11 \div 4 = 7$
- ⬭ $44 \div 22 = 2$
- ⬭ $44 \div 4 = 11$

8 Which number sentence is NOT included in the same fact family as $20 \div 10 = 2$? Mark your answer.

- ⬭ $2 \times 10 = 20$
- ⬭ $10 \times 2 = 20$
- ⬭ $4 \times 10 = 40$
- ⬭ $20 \div 2 = 10$

9 Which multiplication sentence is related to $18 \div 3 = 6$? Mark your answer.

- ⬭ $3 \times 3 = 9$
- ⬭ $6 \times 3 = 18$
- ⬭ $6 \times 2 = 12$
- ⬭ $3 \times 4 = 12$

10 What is a multiplication sentence that is related to $12 \div 6 = 2$?

_____

11 What is a division sentence that is related to $2 \times 11 = 22$?

_____

12 What is a multiplication sentence that is related to $28 \div 7 = 4$?

_____

3.7A generate a table of paired numbers based on a real-life situation such as insects and legs

1 The table shows the cost of sandwiches at George's Sandwich Shop. How much do 6, 8, and 10 sandwiches cost? Mark your answer.

Sandwiches	2	4	6	8	10
Cost	$4	$8	☐	☐	☐

- ⊂ $12, $16, $20
- ⊂ $10, $15, $20
- ⊂ $20, $30, $40
- ⊂ $11, $12, $13

2 The table shows the cost of concert tickets at the Hogg Auditorium in Austin, Texas. Which of the following is a rule for the table? Mark your answer.

Tickets	3	5	7	9	11
Cost	$30	$50	$70	$90	$110

- ⊂ Multiply the number of tickets by $20.
- ⊂ Add $10 to the number of tickets.
- ⊂ Multiply the number of tickets by $10.
- ⊂ Add $20 to the number of tickets.

3 The table shows the cost of CDs at Bob's Discount Music. What is the rule for the table?

CDs	2	4	6	8	10
Cost	$12	$24	$36	$48	$60

_____

_____

4 The table shows the cost of ice cream cones at the Frosty Ice Cream Shop. How much do 6, 8, and 10 ice cream cones cost?

Ice Cream Cones	2	4	6	8	10
Cost	$6	$12	☐	☐	☐

_____

© Harcourt

5 The table shows the cost of cookies at Lou's Cookie Shop. Which of the following is a rule for the table? Mark your answer.

Cookies	5	10	15	20	25
Cost	$10	$20	$30	$40	$50

- ○ Multiply the number of cookies by $10.
- ○ Multiply the number of cookies by $5.
- ○ Multiply the number of cookies by $2.
- ○ Add $2 to the number of cookies.

6 The table shows the cost of comic books at Joe's Comic Shop. How much do 7, 9, and 11 comic books cost? Mark your answer.

Comic books	3	5	7	9	11
Cost	$6	$10	☐	☐	☐

- ○ $12, $16, $20
- ○ $11, $12, $13
- ○ $12, $16, $20
- ○ $14, $18, $22

7 The table shows the cost of dolls at The Doll Boutique. What is the rule for the table?

Dolls	4	6	8	10	12
Cost	$20	$30	$40	$50	$60

_____

_____

8 The table shows the cost of notebooks at Nia's Book Shop. How much do 5, 6, and 7 notebooks cost?

Notebooks	3	4	5	6	7
Cost	$15	$20	☐	☐	☐

_____

3.7B identify and describe patterns in a table of related number pairs based on a meaningful problem and extend the table

1 The table shows the cost of a bottle of shampoo at Good Day Grocery. Find the cost of 8 bottles of shampoo. Mark your answer.

Bottles	3	5	7	8	9	11
Cost	$6	$10	$14	☐	$18	$22

- ⬭ $16
- ⬭ $18
- ⬭ $24
- ⬭ $36

3 The table shows the cost of a rose at Tali's Flower Shop. Find the cost of 9 roses.

Roses	4	6	8	9	10	12
Cost	$12	$18	$24	☐	$30	$36

_____

2 The table shows the cost of a slice of pizza at Roy's Pizza. Find the cost of 12 slices of pizza. Mark your answer.

Slices	2	4	6	8	10	12
Cost	$6	$12	$18	$24	$30	☐

- ⬭ $31
- ⬭ $32
- ⬭ $36
- ⬭ $45

4 The table shows the cost of a poster at Ravi's Poster Store. What is the cost of 10 posters?

Posters	5	6	7	8	9	10
Cost	$25	$30	$35	$40	$45	☐

_____

5 The table shows the cost of a loaf of bread at the Good Bakery. Which is the cost of 12 loaves of bread? Mark your answer.

Loaves	2	4	6	8	10	12
Cost	$8	$16	$24	$32	$40	☐

- ⊂⊃ $41
- ⊂⊃ $48
- ⊂⊃ $42
- ⊂⊃ $35

7 The table shows the cost of a bag of jelly beans at Yasha's Candy Store. What is the cost of 9 bags of jelly beans?

Bags	4	6	8	9	10	12
Cost	$16	$24	$32	☐	$40	$48

_____

6 The table shows the cost of a burger at Larry's Burger Joint. Which is the cost of 6 hamburgers? Mark your answer.

Burgers	3	5	6	7	9	11
Cost	$18	$30	☐	$42	$54	$66

- ⊂⊃ $36
- ⊂⊃ $48
- ⊂⊃ $72
- ⊂⊃ $78

8 The table shows the cost of a balloon ride at the fair. What is the cost of 12 balloon rides?

Rides	5	7	9	11	12	13
Cost	$25	$35	$45	$55	☐	$65

_____

Name _____

↓ 3.8 The student is expected to identify, classify, and describe two- and three-dimensional geometric figures by their attributes. The student compares two-dimensional figures, three-dimensional figures, or both by their attributes using formal geometry vocabulary.

1 Which of the following CANNOT have a bottom, side, or top view that looks like the figure? Mark your answer.

- ⬭ Sphere
- ⬭ Rectangular prism
- ⬭ Cube
- ⬭ Square pyramid

2 Which shows how many sides the figure has? Mark your answer.

- ⬭ 2
- ⬭ 4
- ⬭ 6
- ⬭ 8

3 What is the correct name for the polygon?

_____

4 How many vertices does the figure have?

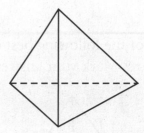

_____

5 How many sides does a pentagon have?

_____

6 Which shows how many faces the figure has? Mark your answer.

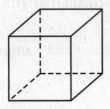

- ⬭ 2
- ⬭ 4
- ⬭ 5
- ⬭ 6

7 Which of the following best describes the figure? Mark your answer.

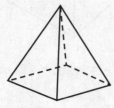

- ⬭ Cone
- ⬭ Cylinder
- ⬭ Square pyramid
- ⬭ Sphere

8 What word correctly name this triangle?

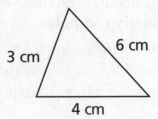

_____

9 How are a cylinder and a sphere alike?

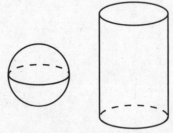

_____

_____

10 What is the name of the figure?

_____

© Harcourt

3.9A   identify congruent two-dimensional figures

1 Which figure is congruent to the figure below? Mark your answer.

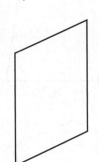

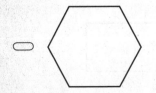

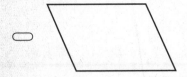

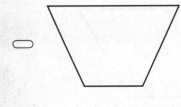

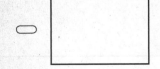

2 Which figure is congruent to the figure below? Write the letter.

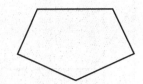

A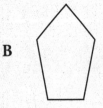

B

C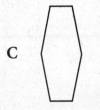

D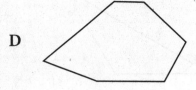

Standards Practice

Practice for TAKS Success

3 Which figure is congruent to the figure
   below? Mark your answer.

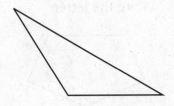

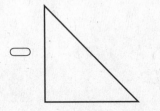

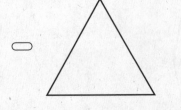

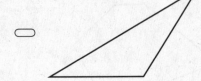

4 Which figure is congruent to the figure
   below? Write the letter.

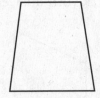

A

B

C

D

_____

Name _____

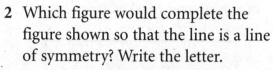

3.9B create two-dimensional figures with lines of symmetry using concrete models and technology

1 Which figure would complete the figure shown so that the line is a line of symmetry? Mark your answer.

2 Which figure would complete the figure shown so that the line is a line of symmetry? Write the letter.

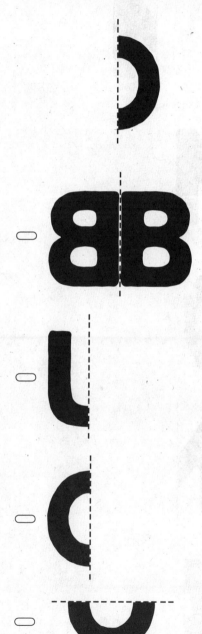

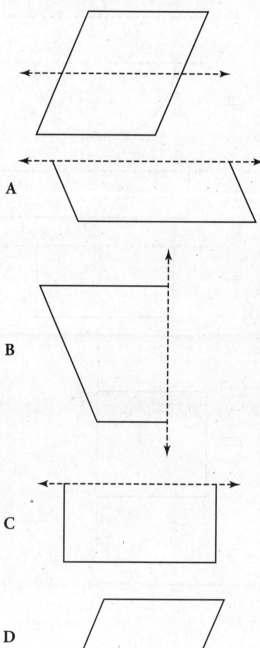

© Harcourt

3 Which figure would complete the figure shown so that the line is a line of symmetry? Mark your answer.

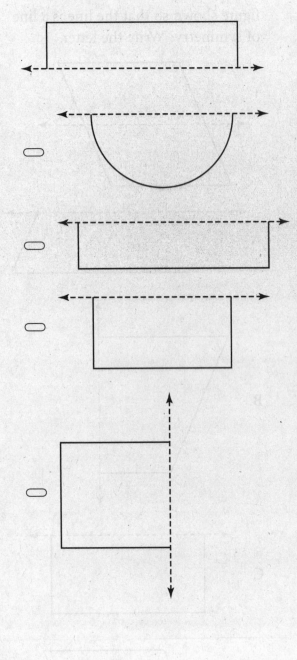

4 Which figure would complete the figure shown so that the line is a line of symmetry? Write the letter.

A

B

C

D

© Harcourt

⭐ 3.9C   identify lines of symmetry in two-dimensional geometric figures

1 Which figure does NOT have a line of symmetry? Mark your answer.

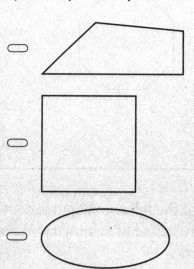

3 How many lines of symmetry does the figure have?

_____

4 Which letter does NOT have any lines of symmetry?

B D X J

_____

2 Which letter has a line of symmetry? Mark your answer.

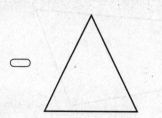

○ P
○ W
○ F
○ G

© Harcourt

5 How many lines of symmetry does the figure have? Mark your answer.

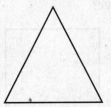

- ⬭ 0
- ⬭ 1
- ⬭ 2
- ⬭ 3

6 Which of the following figures has exactly 1 line of symmetry? Mark your answer.

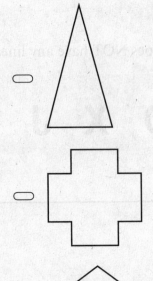
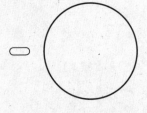

7 How many lines of symmetry does the figure have?

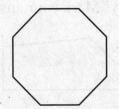

_____

8 Which of the following figures does NOT have a line of symmetry? Write the letter.

A
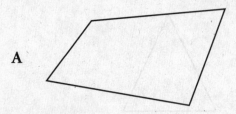

B
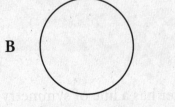

C

D
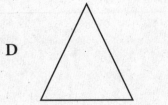

_____

© Harcourt

★ 3.10 The student is expected to locate and name points on a number line using whole numbers and fractions, including halves and fourths.

1 Which number does point *X* best represent on the number line? Mark your answer.

- ⬭ 5
- ⬭ 10
- ⬭ 11
- ⬭ 15

2 Which number is greater than point *A* on the number line? Mark your answer.

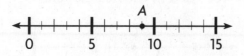

- ⬭ 10
- ⬭ 7
- ⬭ 5
- ⬭ 1

3 Which number does point *Y* best represent on the number line? Mark your answer.

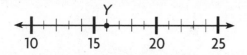

- ⬭ 10
- ⬭ 11
- ⬭ 16
- ⬭ 17

4 Write a number that is less than point *B* but greater than 3 on the number line.

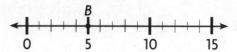

_____

5 What fraction names the point?

_____

6 Which number does point *X* best represent on the number line? Mark your answer.

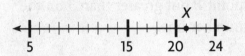

- ⬭ 5
- ⬭ 17
- ⬭ 19
- ⬭ 21

7 Which shows the number represented by point *C* rounded to the nearest ten? Mark your answer.

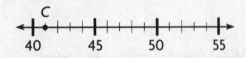

- ⬭ 40
- ⬭ 42
- ⬭ 50
- ⬭ 60

8 Which number is greater than point *D* but less than 55 on the number line? Mark your answer.

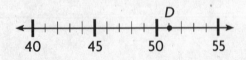

- ⬭ 52
- ⬭ 50
- ⬭ 48
- ⬭ 41

9 What fraction names the point?

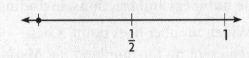

_____

10 Write the number represented by point *D* rounded to the nearest ten.

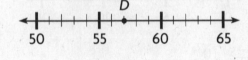

_____

© Harcourt

11 Which number is greater than point *A* on the number line? Mark your answer.

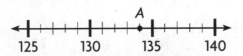

- ⬭ 128
- ⬭ 132
- ⬭ 134
- ⬭ 138

12 Which number does point *Y* best represent on the number line? Mark your answer.

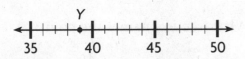

- ⬭ 37
- ⬭ 39
- ⬭ 41
- ⬭ 43

13 Which shows the number represented by point *C* rounded to the nearest ten? Mark your answer.

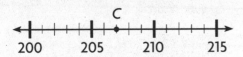

- ⬭ 200
- ⬭ 205
- ⬭ 210
- ⬭ 220

14 What fraction names the point?

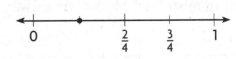

_____

15 Write a number that is greater than point *B* but less than 120 on the number line.

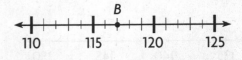

_____

© Harcourt

number is less than point *A* on
number line? Mark your answer.

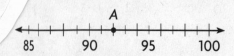

- ○ 97
- ○ 95
- ○ 93
- ○ 89

18 Write a number that is greater than
point *B* but less than 85 on the
number line.

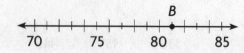

_____

17 Which number does point *Y* best
represent on the number line? Mark
your answer.

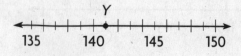

- ○ 137
- ○ 139
- ○ 141
- ○ 143

19 What fraction names the point?

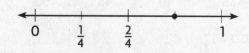

_____

© Harcourt

3.11A use linear measurement tools to estimate and measure lengths using standard units

1 Sandy wants to measure a straw. Which shows about how long the straw is in centimeters? Mark your answer.

⬭ 2 cm
⬭ 20 cm
⬭ 200 cm
⬭ 2,000 cm

2 Jose wants to know how long his kitchen is. Which is the best measure? Mark your answer.

⬭ 75 yards
⬭ 400 feet
⬭ 60 feet
⬭ 6 yards

3 The drive from Roberto's house in Austin to his school is about 3 kilometers. Which shows how many meters that is? Mark your answer.

⬭ 3 meters
⬭ 30 meters
⬭ 300 meters
⬭ 3,000 meters

4 Maria wants to measure the length of her guitar. Which of the following units would be the best to use: miles, yards, or feet?

_____

5 Estimate the length of the spoon to the nearest centimeter.

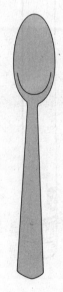

_____

6 Michelle has a diary that is 2 decimeters long. How many centimeters is that? Mark your answer.

- ⊂⊃ 2 centimeters
- ⊂⊃ 6 centimeters
- ⊂⊃ 20 centimeters
- ⊂⊃ 200 centimeters

7 Which is the length of the pencil to the nearest half-inch? Mark your answer.

- ⊂⊃ $3\frac{1}{2}$ inches
- ⊂⊃ 4 inches
- ⊂⊃ $4\frac{1}{2}$ inches
- ⊂⊃ 5 inches

8 Pascale wants to know the distance from his house to his grandmother's house. His grandmother lives in another town. Which of the following units would be best to use: yards, miles, feet, or inches?

_____

9 Richard wants to measure the length of a computer screen. Which of the following could be Richards measurement: 4 in., 14 in., 104 in., or 400 in.?

_____

10 Farrah has a mirror that is 3 decimeters long. How many centimeters is that?

_____

Name _____

3.11B use standard units to find the perimeter of a shape

1 Betty is putting a border around a picture that she drew. The picture is 7 inches tall and 7 inches wide. What is the perimeter of the border? Mark your answer.

7 inches (right side)

7 inches (bottom)

○ 14 inches
○ 27 inches
○ 28 inches
○ 49 inches

2 A rectangle has a perimeter of 20 units. Which shows this rectangle? Mark your answer.

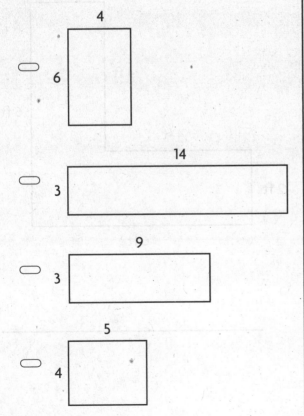

○ 4 (top) / 6 (side)

○ 14 (top) / 3 (side)

○ 9 (top) / 3 (side)

○ 5 (top) / 4 (side)

3 Lars drew a rectangle with two sides that measure 8 inches and two sides that measure 4 inches. What is the perimeter of the rectangle?

8 inches

4 inches

_____

4 Isabella has a swimming pool shaped like a polygon. What is the perimeter of the swimming pool?

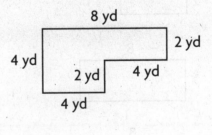

8 yd
2 yd
4 yd
2 yd
4 yd
4 yd

_____

© Harcourt

5 The window is 8 feet long and 3 feet wide. What is the perimeter of the window? Mark your answer.

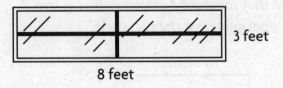

3 feet

8 feet

○ 11 feet
○ 12 feet
○ 22 feet
○ 31 feet

6 A rectangle has a perimeter of 14 units. Which shows this rectangle? Mark your answer.

○   16

4

○   5

2

○   8

3

○   8

2

7 Frida drew a rectangle with two sides that measure 7 centimeters and two sides that measure 3 centimeters. What is the perimeter of the rectangle?

7 cm

3 cm

_____

8 Cheryl has a table shaped like the polygon below. What is the perimeter of the table?

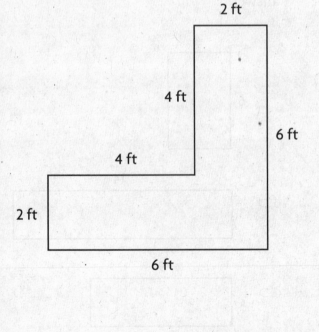

2 ft

4 ft

6 ft

4 ft

2 ft

6 ft

_____

© Harcourt

3.11C use concrete and pictorial models of square units to determine the area of two-dimensional surfaces

1 Tara arranges 1-unit square tiles to form a letter T like below. What is the area of the T? Mark your answer.

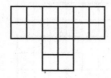

- ⬭ 9 square units
- ⬭ 16 square units
- ⬭ 20 square units
- ⬭ 25 square units

2 Jacinda drew a rectangle with two sides that measure 9 inches and two sides that measure 4 inches. What is the area of the rectangle? Mark your answer.

9 inches

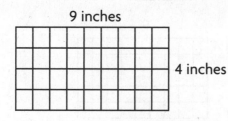

4 inches

- ⬭ area = 13 square units
- ⬭ area = 26 square units
- ⬭ area = 36 square units
- ⬭ area = 48 square units

3 Bruce has a square mirror with side lengths of 7 centimeters. What is the area of the mirror?

7 cm

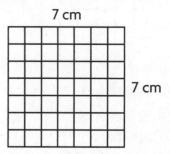

7 cm

4 A rectangle has an area of 32 units. Which shows this rectangle? Write the letter.

4

A   8
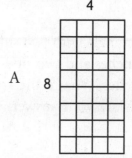

11

B   2
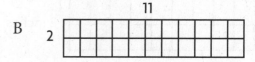

9

C   3
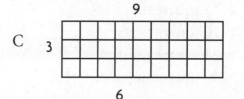

6

D   4
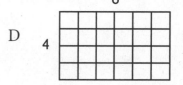

5 Gordon has a kitchen table. The top of the table is a rectangle that is 4 feet by 6 feet. What is the area of the desktop? Mark your answer.

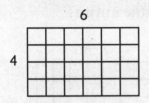

- ⬭ 10 square feet
- ⬭ 22 square feet
- ⬭ 24 square feet
- ⬭ 44 square feet

6 Corbin drew a rectangle with two sides that measure 10 inches and two sides that measure 2 inches. What is the area of the rectangle? Mark your answer.

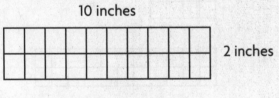

- ⬭ 12 square units
- ⬭ 20 square units
- ⬭ 24 square units
- ⬭ 30 square units

7 Kiefer has a notebook with 2 sides that measure 10 inches and 2 sides that measure 8 inches. What is the area of the notebook?

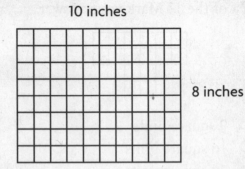

8 A rectangle has an area of 45 units. Which shows this rectangle? Write the letter.

A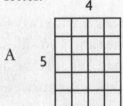

B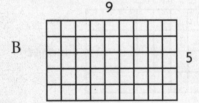

C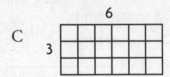

D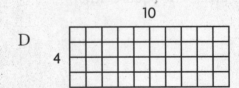

3.11D identify concrete models that approximate standard units of weight/mass and use them to measure weight/mass

1 Karen as a book bag that weighs 48 ounces. Which shows how many pounds are in 48 ounces? Mark your answer.

⬭ 2 pounds
⬭ 3 pounds
⬭ 4 pounds
⬭ 5 pounds

2 Which of the following items would you measure using grams as the unit of measure? Mark your answer.

peanut

boat

dog

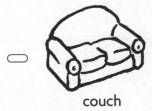
couch

3 Lee has a truck. What unit of measure is best to use to find the weight of the object: pounds or ounces?

_____

4 Padma has a toy that weighs 1 pound. How many ounces are in 1 pound.

_____

5 Hector has a dog that weighs 5,000 grams. How many kilograms are in 5,000 grams?

_____

© Harcourt

6 John has a chair that weighs 6 pounds. Which shows how many ounces are in 6 pounds? Mark your answer.

- ⬭ 6 ounces
- ⬭ 69 ounces
- ⬭ 96 ounces
- ⬭ 100 ounces

7 Vadim has a book that weighs 1 kilogram. Which shows how many grams are in 1 kilogram? Mark your answer.

- ⬭ 250 grams
- ⬭ 500 grams
- ⬭ 1,000 grams
- ⬭ 2,000 grams

8 Kali has a bag of flour. Which of the following is the best unit of measure? Mark your answer.

- ⬭ 2 ounces
- ⬭ 10 ounces
- ⬭ 2 pounds
- ⬭ 200 pounds

9 Louis has a dog that weighs 7,000 grams. How many kilograms equal 7,000 grams?

_____

10 Which of the following items would you measure using pounds as the unit of measure?

cherry

paper clip

nut

person

_____

© Harcourt

3.11E identify concrete models that approximate standard units for capacity and use them to measure capacity

1 Billy wants to measure the capacity of a kiddie pool. Which unit should Billy use to measure? Mark your answer.

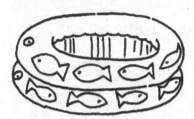

- ⚬ quart
- ⚬ gallon
- ⚬ pint
- ⚬ cup

2 Blake wants to measure the capacity of an eyedropper. What unit should Blake use to measure? Mark your answer.

- ⚬ quart
- ⚬ gallon
- ⚬ milliliter
- ⚬ liter

3 Which of the following items would you measure using liters as the unit of measure?

juice carton

mug

teacup

eyedropper

Name _____

4 Bree wanted to measure the capacity of a milk carton. Which unit should he use to measure? Mark your answer.

milk carton

○ gallon
○ cup
○ milliliter
○ quart

5 Nick wants to measure the capacity of the glass of juice he is drinking. What unit should he use to measure? Mark your answer.

○ quart
○ gallon
○ ounce
○ milliliter

6 Which of the following items would you measure using milliliters as the unit of measure?

bottle cap with liquid

fish tank

gallon of milk

orange juice carton

© Harcourt

3.11F use concrete models that approximate cubic units to determine the volume of a given container or other three-dimensional geometric figure

1 Ahmed has a box in the shape of the figure below. What is the volume of the box? Mark your answer.

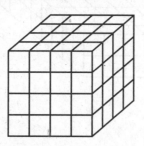

   ◌ 46 cubic units

   ◌ 52 cubic units

   ◌ 64 cubic units

   ◌ 456 cubic units

2 What is the volume of the rectangular prism? Mark your answer.

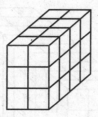

   ◌ 24 cubic units

   ◌ 32 cubic units

   ◌ 75 cubic units

   ◌ 183 cubic units

3 Johann has a box in the shape of the figure. What is the volume of the box?

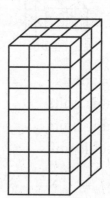

_____

4 What is the volume of the rectangular prism?

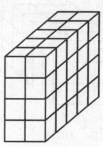

_____

© Harcourt

5 Uma has a box in the shape of the figure. What is the volume of the box? Mark your answer.

◯ 8 cubic units
◯ 12 cubic units
◯ 22 cubic units
◯ 222 cubic units

6 What is the volume of the rectangular prism? Mark your answer.

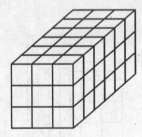

◯ 33 cubic units
◯ 36 cubic units
◯ 54 cubic units
◯ 336 cubic units

7 Kyoko has a box in the shape of the figure. What is the volume of the box?

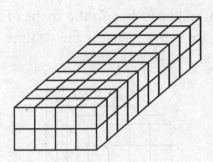

_____

8 What is the volume of the rectangular prism?

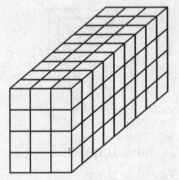

_____

© Harcourt

3.12A **use a thermometer to measure temperature**

1 Luis lives in San Antonio. He looks at a thermometer on the left before leaving in the morning. He looked at the thermometer on the right in the afternoon. How much did the temperature change? Mark your answer.

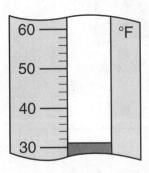 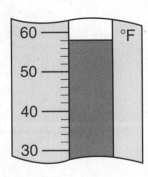

- ⬭ 27°F
- ⬭ 33°F
- ⬭ 37°F
- ⬭ 42°F

2 Juanita lives in Abilene. On April 14th, she looks out her window at her thermometer. What temperature did the thermometer show? Mark your answer.

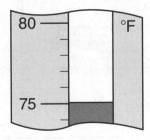

- ⬭ 62°F
- ⬭ 70°F
- ⬭ 75°F
- ⬭ 81°F

3 Ross lives in Tyler. On July 27th, he looks out his window at his thermometer. What temperature did the thermometer show?

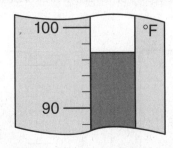

4 Julie is swimming in August. What temperature would be best for swimming: 19°F, 58°F, 85°F, or 185°F?

5 Marissa lives in Lubbock. She looks at a thermometer on the left before leaving in the morning. She looked at the thermometer on the right in the afternoon. How much did the temperature change? Mark your answer.

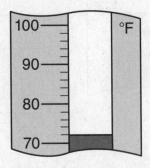 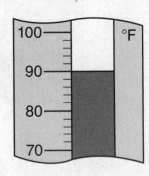

- ⬭ 18°F
- ⬭ 20°F
- ⬭ 21°F
- ⬭ 68°F

6 Igor is skiing in January. What temperature would be best for skiing? Mark your answer.

- ⬭ 26°F
- ⬭ 46°F
- ⬭ 77°F
- ⬭ 98°F

7 Kirby lives in Dallas. He looks out his window at his thermometer. What temperature did the thermometer show?

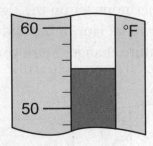

_____

8 Bianca lives in Plainview. She looks out her window at her thermometer. What temperature did the thermometer show?

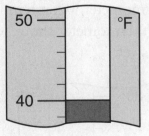

_____

© Harcourt

3.12B **tell and write time shown on analog and digital clocks**

1 Kevin ran in the Texas Half Marathon. His total time is shown on the clock below. What was his time for the race? Mark your answer.

○ 2 hours, 48 minutes, 57 seconds
○ 2 hours, 53 minutes, 48 seconds
○ 2 hours, 57 minutes, 48 seconds
○ 2 hours, 57 minutes, 52 seconds

2 What time is shown on the clock? Mark your answer.

○ 1:45
○ 2:55
○ 3:55
○ 11:05

3 Kirsten's flight leaves at a quarter past 8 in the evening from Dallas International Airport in Dallas, Texas. What is the time?

_____

4 Billy is going to a movie at his local movie theater in Dallas, Texas. The film is showing at fifteen minutes to nine. What is the time?

_____

5 What time is shown on the clock?

_____

6 Tera meets a friend at the park at the time shown on the clock. What time is shown on the clock? Mark your answer.

- ⬭ 4:45
- ⬭ 9:45
- ⬭ 10:15
- ⬭ 11:45

7 Cindy leaves the house at half past eight in the morning. Which shows the time Cindy left?

- ⬭ 8:30 P.M.
- ⬭ 9:00 P.M.
- ⬭ 8:30 A.M.
- ⬭ 9:00 A.M.

8 Which of the times do most third-grade students start their day at school? Mark your answer.

- ⬭ 5:00 A.M.
- ⬭ 8:00 A.M.
- ⬭ 2:30 P.M.
- ⬭ 8:00 P.M.

9 What time is shown on the clock? Write your answer in words.

_____

10 What time is a quarter till nine?

_____

11 Henry meets a friend at the library at the time shown on the clock. What time is shown on the clock?

_____

3.13A collect, organize, record, and display data in pictographs and bar graphs where each picture or cell might represent more than one piece of data

1 Victor wanted to make a bar graph to show how many CDs he has. Below is the information he will graph. Which type of CD has the tallest bar? Mark your answer.

- ⬭ 16 rock CDs
- ⬭ 19 pop CDs
- ⬭ 11 alternative CDs
- ⬭ 22 country CDs

3 William made a bar graph to show how many books he has. Which two types of books have a total of 42 books?

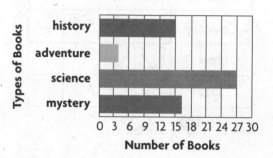

2 Which scale would be the best scale to use if you wanted to make a bar graph of the data shown in the pictograph?

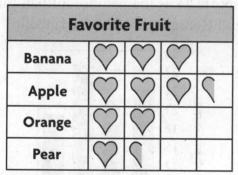

Favorite Fruit

key: Each ♥ = 20 votes

- ⬭ 0, 5, 10, 15, 20, 25, 30
- ⬭ 0, 2, 4, 6, 8, 10, 12
- ⬭ 0, 10, 20, 30, 40, 50, 60
- ⬭ 0, 20, 40, 60, 80, 100

4 Melissa made a bar graph to show how many stickers she has. What type of sticker has the tallest bar?

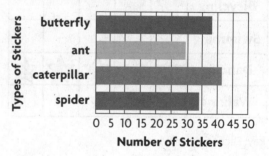

© Harcourt

5 Which scale would be the best scale to use if you wanted to make a bar graph of the data shown in the pictograph?

Favorite Music				
Hip Hop	♡	♡		
Country	♡	♡	♡	◗
Pop	♡	♡	◗	
Jazz	♡	◗		

key: Each ♡ = 8 votes

○ 0, 5, 10, 15, 20, 25, 30
○ 0, 8, 16, 24, 32, 40, 48
○ 0, 10, 20, 30, 40, 50, 60
○ 0, 20, 40, 60, 80, 100

6 Which scale would be the best scale to use if you wanted to make a bar graph of the data shown in the pictograph?

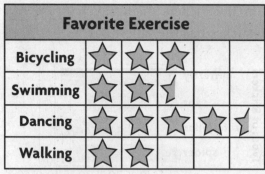

Favorite Exercise					
Bicycling	☆	☆	☆		
Swimming	☆	☆	◖		
Dancing	☆	☆	☆	☆	◖
Walking	☆	☆			

key: Each ☆ = 8 votes

○ 0, 6, 12, 18, 24, 30, 34
○ 0, 2, 4, 6, 8, 10, 12
○ 0, 10, 20, 30, 40, 50, 60
○ 0, 8, 16, 24, 32, 40, 48

7 Danielle made a bar graph to show how many books she has read. How many total books has Danielle read?

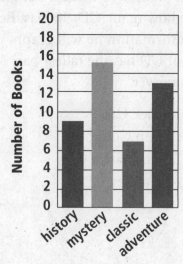

8 Georgia made a bar graph to show how many games she played in several sports. How many more softball games did Georgia play than basketball?

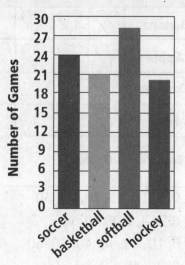

Practice for TAKS Success   116   **Standards Practice**

© Harcourt

3.13B interpret information from pictographs and bar graphs

1 The bar graph shows the results of Dylan's survey of favorite foods. Which food received more votes than hamburgers? Mark your answer.

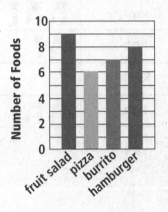

- ⬭ Fruit Salad
- ⬭ Pizza
- ⬭ Burrito
- ⬭ Hamburger

2 Victoria made a pictograph to show how many books she has. How many novels books does Victoria have? Mark your answer.

Types of Books	
Humor	📕
Biography	📙
Novels	📕 📙
Science	📘 📗

key: Each 📕 = 6 Books

- ⬭ 3   ⬭ 9
- ⬭ 6   ⬭ 12

3 Evan made a pictograph to show how many CDs he has. How many rock CDs does Evan have?

Types of CDs	
Pop	◎ ◎
World	◎ ◗
Country	◗
Rock	◎ ◎ ◎

key: Each ◎ = 10 CDs

4 The bar graph shows the results of Ruth's survey of her friends' favorite Texas natural landmarks. Which landmark received about 5 more votes than Catfish Creek?

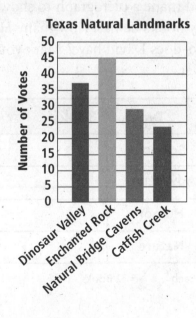

© Harcourt

5 The bar graph shows the results of Doris's survey of favorite foods. Which food received 4 more votes than chicken-fried steak? Mark your answer.

Favorite Foods

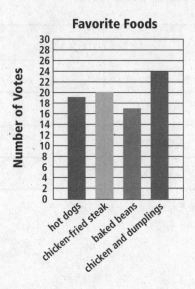

- ⊂⊃ Baked beans
- ⊂⊃ Chicken-fried steak
- ⊂⊃ Hot Dogs
- ⊂⊃ Chicken and dumplings

6 David made a pictograph to show how many books he has. How many Humor books does David have? Mark your answer.

Types of Books	
Science	📕
Adventure	📗
Humor	📕 📗
Nature	📗 📗

key: Each 📗 = 12 Books

- ⊂⊃ 6   ⊂⊃ 18
- ⊂⊃ 12   ⊂⊃ 24

7 Nelson made a pictograph to show how many mp3s he has. How many Country mp3s does Nelson have?

Types of mp3s	
Country	♫♫♪
Pop	♫♪
Rock	♪
Jazz	♫♫

key: Each ♫ = 12 mp3s

_____

8 The bar graph shows the results of Billy's survey of favorite games. Which two games received 35 votes?

Favorite Games

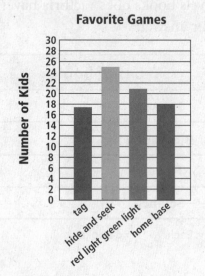

_____

© Harcourt

★ 3.13C **use data to describe events as more likely than, less likely than, or equally likely as**

1 Jolene is going to pull a piece of fruit out of a bag. What is the probability of pulling an apple out of the bag? Mark your answer.

○ Impossible
○ Unlikely
○ Likely
○ Certain

3 What is the probability of pulling a tile marked "B" out of the bag: impossible, unlikely, likely, or certain?

_____

2 What is the probability of the spinner landing on the letter A, B, or C? Mark your answer.

○ Impossible
○ Unlikely
○ Likely
○ Certain

4 What is the probability of pulling a white marble out of the bag: impossible, unlikely, likely, or certain?

_____

5 Kim is going to pull a jellybean out of a bag. What is the probability of pulling a black jellybean out of the bag? Mark your answer.

- ⬭ Impossible
- ⬭ Unlikely
- ⬭ Likely
- ⬭ Certain

6 Which of the following best describes the probability of landing on a white or a striped section of the spinner? Mark your answer.

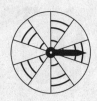

- ⬭ Impossible
- ⬭ Unlikely
- ⬭ Likely
- ⬭ Certain

7 What is the probability of the spinner landing on the letter A: impossible, unlikely, likely, or certain?

_____

8 What is the probability of pulling a white marble from the bag: impossible, unlikely, likely, or certain?

_____

© Harcourt

Record your answer in the boxes below each question. Then fill in the bubbles. Be sure to use the correct place value.

1 **3.1C** What is the missing number? Mark your answer.

1 quarter + ☐ nickels = $1.00

2 **3.5B** Dallas is 61 miles from Canton. Canton is 12 miles from Colfax. Round to the nearest ten to estimate the total distance from Dallas to Canton then to Colfax. Mark your answer.

3 **3.4B** Timmy and his father go bird watching at Bentsen-Rio Grande Valley State Park in Brownsville. The park charges a $5 entrance fee per person. How much did they pay to go in? Mark your answer.

4 **3.10** What number is one greater than point X on the number line? Mark your answer.

55   85

5 🔻 **3.4C** A group of 84 students have a water volleyball tournament at Ray Roberts Lake. Each team has 7 players. How many teams are there? Mark your answer.

7 🔻 **3.4C** Lisa has 50 stamps in her stamp book. Each page has 5 stamps. How many pages are in Lisa's stamp book? Mark your answer.

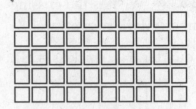

6 🔻 **3.3B** What is the missing number? Mark your answer.

$$99 - \square = 49$$

8 🔻 **3.4B** Cindy's class goes on a trip to the Museum of Nature and Science in Dallas, Texas. The museum tickets cost $6 for each of the 12 students. What was the total cost the students paid? Mark your answer.

9 🔻 3.4B Bob bought 7 boxes of colored pencils for his store. Each box contains 13 packages. How many total packages did Bob buy? Mark your answer.

11 🔻 3.13B Brian made a tally table to record his friends' votes for their favorite Texas animal. How many of Terry's friends voted for armadillo as their favorite Texas animal? Mark your answer.

Favorite Animal	
Animal	Tally
Armadillo	卌 卌 III
Mockingbird	卌 卌 卌 I
Texas Longhorn	卌 卌 II
Bat	卌 卌 I

10 🔻 3.3B Sarah asked 200 of her classmates to name their favorite bird. Of the classmates asked, 136 chose mockingbirds. How many of the classmates chose a different bird? Mark your answer.

12 🔻 3.4B What is the missing factor? Mark your answer.

$$8 \times 12 = \square \times 8$$

© Harcourt

13 **3.4C** What is the missing number? Mark your answer.

$$96 \div \square = 8$$

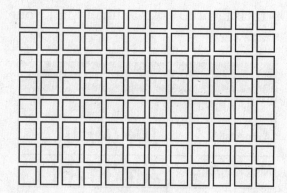

14 **3.3B** Cathy asked 250 of her classmates, to list their favorite Texas flower. Of the classmates asked, 179 chose bluebonnets. How many of the classmates chose a different flower than bluebonnet? Mark your answer.

15 **3.4B** Gerald walks 3 miles each day for 5 days. How many total miles does Gerald walk? Mark your answer.

16 **3.4C** What is the missing number? Mark your answer.

$$\square \times 12 = 24$$

17 3.4C Allie bought 60 pencils. The pencils came in packages of 6. How many packages did Allie buy? Mark your answer.

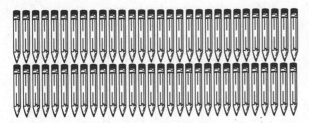

⓪	⓪
①	①
②	②
③	③
④	④
⑤	⑤
⑥	⑥
⑦	⑦
⑧	⑧
⑨	⑨

18 3.6A What is the next number in the pattern? Mark your answer.

72, 66, 60, 54, 48, ☐

⓪	⓪
①	①
②	②
③	③
④	④
⑤	⑤
⑥	⑥
⑦	⑦
⑧	⑧
⑨	⑨

19 3.4B A group of 5 children buy tickets for a movie. Each ticket cost $9. How much total money did the children spend? Mark your answer.

⓪	⓪
①	①
②	②
③	③
④	④
⑤	⑤
⑥	⑥
⑦	⑦
⑧	⑧
⑨	⑨

20 3.4B Keith's bakery buys 7 dozen eggs each day. There are 12 eggs in a dozen. How many total eggs does the bakery buy each day? Mark your answer.

⓪	⓪
①	①
②	②
③	③
④	④
⑤	⑤
⑥	⑥
⑦	⑦
⑧	⑧
⑨	⑨

21 🔻 **3.4C** A group of 27 students are playing a game in gym class. They want to form equal teams with 9 students on each team. How many total teams can they form? Mark your answer.

⓪	⓪
①	①
②	②
③	③
④	④
⑤	⑤
⑥	⑥
⑦	⑦
⑧	⑧
⑨	⑨

22 🔻 **3.3B** Lilly read two books this week. The first book had 52 pages. The second book had 45 pages. How many total pages did Lilly read? Mark your answer.

⓪	⓪
①	①
②	②
③	③
④	④
⑤	⑤
⑥	⑥
⑦	⑦
⑧	⑧
⑨	⑨

23 🔻 **3.5A** In 2000, Domino, Texas had a population of 52. What is the population rounded to the nearest ten? Mark your answer.

⓪	⓪
①	①
②	②
③	③
④	④
⑤	⑤
⑥	⑥
⑦	⑦
⑧	⑧
⑨	⑨

24 🔻 **3.4C** Tina spends $11 on paintbrushes. Each brush costs $1. How many brushes did Tina buy? Mark your answer.

⓪	⓪
①	①
②	②
③	③
④	④
⑤	⑤
⑥	⑥
⑦	⑦
⑧	⑧
⑨	⑨

© Harcourt

25 🔻 **3.4B** What is the missing factor? Mark your answer.

$$8 \times \square = 64$$

27 🔻 **3.6A** Brett keeps track of the number of miles that he bicycles each week. How many miles would Brett bicycle the next week, if the pattern continues? Mark your answer.

$$21, 28, 35, 42, \square$$

26 🔻 **3.11E** Kathy has a piece of ribbon that is 4 decimeters long. How long is the ribbon in centimeters? Mark your answer.

28 🔻 **3.4B** Elsie has 11 packages of trading cards. Each package contains 8 cards. How many total cards does Elsie have? Mark your answer.

© Harcourt

29 ♦ 3.5A   The driving distance from Livingston to Houston is about 74 miles. What is 74 rounded to the nearest ten? Mark your answer.

⓪	⓪
①	①
②	②
③	③
④	④
⑤	⑤
⑥	⑥
⑦	⑦
⑧	⑧
⑨	⑨

31 ♦ 3.3B   Which number is ten greater than point M on the number line? Mark your answer.

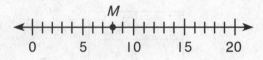

⓪	⓪
①	①
②	②
③	③
④	④
⑤	⑤
⑥	⑥
⑦	⑦
⑧	⑧
⑨	⑨

30 ♦ 3.4B   Leanne has 4 sets of paints. Each set has 5 tubes of different colors. How many total tubes does Leanne have? Mark your answer.

⓪	⓪
①	①
②	②
③	③
④	④
⑤	⑤
⑥	⑥
⑦	⑦
⑧	⑧
⑨	⑨

32 ♦ 3.4B   Martha has 9 packages of markers. Each package has 6 markers. How many total markers does Martha have? Mark your answer.

⓪	⓪
①	①
②	②
③	③
④	④
⑤	⑤
⑥	⑥
⑦	⑦
⑧	⑧
⑨	⑨

© Harcourt

33 3.4B Students order 6 pizzas. Each pizza has 6 slices. What is the total number of slices of pizza? Mark your answer.

⓪	⓪
①	①
②	②
③	③
④	④
⑤	⑤
⑥	⑥
⑦	⑦
⑧	⑧
⑨	⑨

34 3.3B Jasmine's bakery sold 25 dinner rolls on Friday and 73 rolls on Saturday. How many total rolls did Jasmine sell in two days? Mark your answer.

⓪	⓪
①	①
②	②
③	③
④	④
⑤	⑤
⑥	⑥
⑦	⑦
⑧	⑧
⑨	⑨

35 3.3B In 2000, the population of Impact, Texas was 39 and the population of Mustang, Texas was 47. What was the total population of both towns? Mark your answer.

⓪	⓪
①	①
②	②
③	③
④	④
⑤	⑤
⑥	⑥
⑦	⑦
⑧	⑧
⑨	⑨

36 3.5A Danny's uncle is reading a book that has 78 pages. What is 78 rounded to the nearest ten? Mark your answer.

⓪	⓪
①	①
②	②
③	③
④	④
⑤	⑤
⑥	⑥
⑦	⑦
⑧	⑧
⑨	⑨

© Harcourt

37 🔻 **3.5A**   The drive from Paris, Texas to Dallas is 102 miles. The drive from Paris to Bullard, Texas is 137 miles. Round to the nearest ten how much longer the trip to Bullard is. Mark your answer.

⓪	⓪
①	①
②	②
③	③
④	④
⑤	⑤
⑥	⑥
⑦	⑦
⑧	⑧
⑨	⑨

39 🔻 **3.4B**   An art exhibit at the Contemporary Arts Museum of Houston had a photography section with 4 rows and 9 photographs in each row. How many total photographs were in the exhibit? Mark your answer.

⓪	⓪
①	①
②	②
③	③
④	④
⑤	⑤
⑥	⑥
⑦	⑦
⑧	⑧
⑨	⑨

38 🔻 **3.4A**   What number belongs in the box? Mark your answer.

$$5 \times 8 = 4 \times \square$$

⓪	⓪
①	①
②	②
③	③
④	④
⑤	⑤
⑥	⑥
⑦	⑦
⑧	⑧
⑨	⑨

40 🔻 **3.4B**   Cindy buys 3 postcards at the Houston Museum of Natural Science. Each postcard costs 10¢. What was the total amount Cindy paid for all the postcards? Mark your answer.

⓪	⓪
①	①
②	②
③	③
④	④
⑤	⑤
⑥	⑥
⑦	⑦
⑧	⑧
⑨	⑨

41 🔻 **3.4C**   A group of students paid a total of $48 to buy tickets for a movie. Each ticket cost $8. How many students were in the group? Mark your answer.

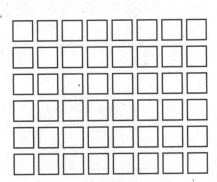

42 🔻 **3.10**   What is the number represented by point *H* rounded to the nearest ten? Mark your answer.

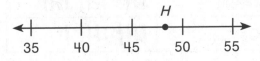

43 🔻 **3.3A**   What number makes this number sentence true? Mark your answer.

$$9 + \square = 98$$

44 🔻 **3.4C**   Krista has 24 baseball cards. The cards come in packages of 6 cards each. How many packages does Krista have? Mark your answer.

45 🔻 3.4B   What is the missing factor? Mark your answer.

$$(3 \times 4) \times 5 = \square \times 20$$

⓪	⓪
①	①
②	②
③	③
④	④
⑤	⑤
⑥	⑥
⑦	⑦
⑧	⑧
⑨	⑨

46 🔻 3.1C   Christina finds 4 pennies, 3 nickels, and 2 dimes in her pocket. What was the total amount of money Christina found? Mark your answer.

⓪	⓪
①	①
②	②
③	③
④	④
⑤	⑤
⑥	⑥
⑦	⑦
⑧	⑧
⑨	⑨

47 🔻 3.4B   Zack jogs for 5 miles each day for 5 days. How many total miles did Zack jog in 5 days? Mark your answer.

⓪	⓪
①	①
②	②
③	③
④	④
⑤	⑤
⑥	⑥
⑦	⑦
⑧	⑧
⑨	⑨

48 🔻 3.13B   Nancy made a tally table to record her friends' votes for their favorite Texas tree. How many of her friends voted for Magnolia as their favorite Texas tree? Mark your answer.

Favorite Tree	
Tree	**Tally**
Dogwood	卌 卌
Magnolia	卌 卌 I
Pecan	卌 卌 IIII
Oak	卌 卌 II

⓪	⓪
①	①
②	②
③	③
④	④
⑤	⑤
⑥	⑥
⑦	⑦
⑧	⑧
⑨	⑨

© Harcourt

49 3.6A What is the next number in the pattern? Mark your answer.

132, 121, 110, 99, ☐

```
0  0
1  1
2  2
3  3
4  4
5  5
6  6
7  7
8  8
9  9
```

51 3.3B The Dallas Cowboys have been champions 5 times. The Houston Comets have been champions 4 times and the San Antonio Spurs have been champions 2 times. How many championships have these Texas teams won altogether? Mark your answer.

```
0  0
1  1
2  2
3  3
4  4
5  5
6  6
7  7
8  8
9  9
```

50 3.4B Mary bought 2 boxes of stickers. Each box contains 6 packages. Each package has 6 stickers. How many total stickers did Mary buy? Mark your answer.

```
0  0
1  1
2  2
3  3
4  4
5  5
6  6
7  7
8  8
9  9
```

52 3.4B What is the missing factor? Mark your answer.

☐ × 12 = 48

```
0  0
1  1
2  2
3  3
4  4
5  5
6  6
7  7
8  8
9  9
```

53 **3.4C**   Frank rides his bicycle 72 miles in one week. He rides the same number of miles per day. He bicycles 6 days in one week. How many miles does Frank ride per day? Mark your answer.

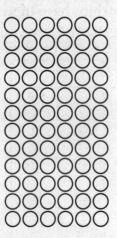

54 **3.4B**   What is the missing factor? Mark your answer.

$$(2 \times 7) \times 3 = 2 \times \square$$

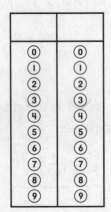

55 **3.3B**   The Austin Museum of Art in Austin, Texas, had 1,457 visitors on Saturday. The center had 1,388 visitors on Sunday. How many more visitors did the center have on Saturday? Mark your answer.

56 **3.1A**   Sawtooth Mountain in western Texas is 7,7<u>4</u>8 feet above sea level. What is the value of the underlined digit? Mark your answer.

57 3.4C   What is the missing number? Mark your answer.

$$\square \div 5 = 3$$

⓪	⓪
①	①
②	②
③	③
④	④
⑤	⑤
⑥	⑥
⑦	⑦
⑧	⑧
⑨	⑨

58 3.4B   Trisha and her mother go bird watching at Bolivar Flats Bird Sanctuary near Galveston, Texas. The park charges a $7 entrance fee per person. How much did they pay to go in? Mark your answer.

⓪	⓪
①	①
②	②
③	③
④	④
⑤	⑤
⑥	⑥
⑦	⑦
⑧	⑧
⑨	⑨

59 3.3B   The El Paso Museum of Art in El Paso, Texas had 568 visitors on Saturday. The center had 611 visitors on Sunday. How many more visitors did the center have on Sunday? Mark your answer.

⓪	⓪
①	①
②	②
③	③
④	④
⑤	⑤
⑥	⑥
⑦	⑦
⑧	⑧
⑨	⑨

60 3.4B   What is the missing factor? Mark your answer.

$$6 \times 11 = \square \times 6$$

⓪	⓪
①	①
②	②
③	③
④	④
⑤	⑤
⑥	⑥
⑦	⑦
⑧	⑧
⑨	⑨

61 🔻 **3.10** What number does point *S* best represent on the number line? Mark your answer.

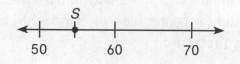

63 🔻 **3.13B** Julie made a pictograph to show how many books she has. How many mystery books does Julie have? Mark your answer.

Types of Books	
Humor	
Nature	
History	
Mystery	
Each = 10 Books	

62 🔻 **3.4B** What is the missing factor? Mark your answer.

$$8 \times \square = 88$$

64 🔻 **3.4B** Melvin bought 7 packages of gum. Each package contains 5 pieces of gum. How many pieces of gum did Melvin buy? Mark your answer.

© Harcourt

65 3.4C A group of 70 students went on a trip to the Carter Museum in Fort Worth, Texas. The students toured the museum in groups of 7. How many groups toured the museum? Mark your answer.

```
□□□□□□□□□□
□□□□□□□□□□
□□□□□□□□□□
□□□□□□□□□□
□□□□□□□□□□
□□□□□□□□□□
□□□□□□□□□□
```

⓪	⓪
①	①
②	②
③	③
④	④
⑤	⑤
⑥	⑥
⑦	⑦
⑧	⑧
⑨	⑨

66 3.3B Richard read a book that had 42 pages and then he read a book that had 15 pages. How many more pages did the first book have? Mark your answer.

⓪	⓪
①	①
②	②
③	③
④	④
⑤	⑤
⑥	⑥
⑦	⑦
⑧	⑧
⑨	⑨

67 3.11E Lisa has a container that holds 2 gallons. How many quarts are equal to 2 gallons.

⓪	⓪
①	①
②	②
③	③
④	④
⑤	⑤
⑥	⑥
⑦	⑦
⑧	⑧
⑨	⑨

68 3.4B A group of students have a basketball tournament. Each team has 8 players and there are 6 teams. How many players are there? Mark your answer.

⓪	⓪
①	①
②	②
③	③
④	④
⑤	⑤
⑥	⑥
⑦	⑦
⑧	⑧
⑨	⑨

69 3.3B Jason has 45 stickers. He uses 30 of the stickers to decorate his notebooks. How many stickers does Jason have left? Mark your answer.

⓪	⓪
①	①
②	②
③	③
④	④
⑤	⑤
⑥	⑥
⑦	⑦
⑧	⑧
⑨	⑨

70 3.4B What is the value of the missing number? Mark your answer.

$$\square \times 9 = 81$$

⓪	⓪
①	①
②	②
③	③
④	④
⑤	⑤
⑥	⑥
⑦	⑦
⑧	⑧
⑨	⑨

71 3.11A Matt has a book that is 2 decimeters wide. How many centimeters is the book? Mark your answer.

⓪	⓪
①	①
②	②
③	③
④	④
⑤	⑤
⑥	⑥
⑦	⑦
⑧	⑧
⑨	⑨

72 3.4C Bob works part time at a diner in El Paso, Texas. He is paid $9 per hour. He makes $108 in one week of work. How many hours does Bob work? Mark your answer.

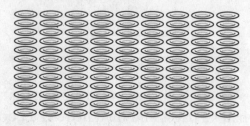

⓪	⓪
①	①
②	②
③	③
④	④
⑤	⑤
⑥	⑥
⑦	⑦
⑧	⑧
⑨	⑨

73 3.4B What is the missing number? Mark your answer.

$$11 \times \square = 110$$

⓪	⓪
①	①
②	②
③	③
④	④
⑤	⑤
⑥	⑥
⑦	⑦
⑧	⑧
⑨	⑨

74 3.11A Estimate the length of the fork to the nearest centimeter. What is the best estimate 2, 3, 8, or 20? Mark your answer.

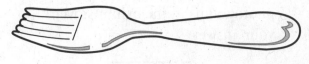

⓪	⓪
①	①
②	②
③	③
④	④
⑤	⑤
⑥	⑥
⑦	⑦
⑧	⑧
⑨	⑨

75 3.4C Kate has 144 baseball cards. The cards come in packages of 12 cards. How many packages does Kate have? Mark your answer.

⓪	⓪
①	①
②	②
③	③
④	④
⑤	⑤
⑥	⑥
⑦	⑦
⑧	⑧
⑨	⑨

© Harcourt

Name _____

76 ⬆ 3.11A The drive from Kate's house to her school is about 20,000 meters. How many kilometers is that? Mark your answer.

77 ⬆ 3.4B Gabby and Rafael visit Meridian State Park near Waco. The park charges an $8 entrance fee per person. What was the total amount Gabby and Rafael paid in fees? Mark your answer.

78 ⬆ 3.3B In 2002, the population of Rocky Mound, Texas was 93 and the population of Impact, Texas was 39. How many more people lived in Rocky Mound than in Impact? Mark your answer.

79 ⬆ 3.4B An art exhibit at the International Museum of Cultures in Dallas had a photography section with 8 rows and 9 photographs in each row. What was the total number of photographs in the exhibits? Mark your answer.

© Harcourt

	A	B	C	D	E	F	G	H
I	32 + 50	542 + 675	29 + 70	4 + 3	59 + 30	132 + 6	10 + 70	8 + 8
J	20 + 14	67 + 10	642 + 76	83 + 10	981 + 452	189 + 3	72 + 10	39 + 60
K	725 + 458	834 + 73	13 + 70	20 + 43	182 + 52	15 + 30	287 + 3	65 + 10
L	17 + 50	687 + 4	37 + 40	97 + 6	20 + 25	2 + 5	69 + 10	975 + 37
M	87 + 6	60 + 12	18 + 80	341 + 852	19 + 50	6 + 9	45 + 20	61 + 30
N	28 + 50	47 + 40	948 + 385	30 + 12	2 + 8	80 + 18	36 + 10	9 + 8
O	60 + 22	125 + 37	17 + 40	9 + 4	45 + 10	347 + 287	37 + 20	187 + 58
P	268 + 71	57 + 30	875 + 95	88 + 10	142 + 22	26 + 70	3 + 9	40 + 19
Q	66 + 10	672 + 394	12 + 10	137 + 51	10 + 15	30 + 20	434 + 363	38 + 30
R	49 + 20	287 + 91	22 + 40	253 + 318	40 + 28	361 + 27	20 + 35	10 + 15

	A	B	C	D	E	F	G	H
I	342 − 132	31 − 10	628 − 54	72 − 40	857 − 254	52 − 40	574 − 527	67 − 40
J	82 − 70	746 − 675	72 − 50	329 − 38	56 − 40	18 − 10	99 − 80	276 − 175
K	5 − 0	51 − 30	32 − 10	82 − 3	6 − 4	69 − 30	199 − 138	85 − 80
L	99 − 90	75 − 60	8 − 4	87 − 70	78 − 30	465 − 28	38 − 10	186 − 144
M	554 − 248	63 − 50	287 − 69	99 − 50	648 − 86	82 − 70	487 − 248	70 − 50
N	176 − 8	29 − 20	33 − 10	223 − 176	91 − 50	75 − 10	8 − 2	30 − 20
O	91 − 80	937 − 19	869 − 753	59 − 10	66 − 20	564 − 442	85 − 30	485 − 378
P	338 − 78	85 − 60	145 − 9	65 − 20	8 − 6	75 − 50	273 − 152	55 − 20
Q	75 − 20	31 − 20	58 − 6	998 − 995	48 − 30	113 − 108	83 − 50	678 − 518
R	82 − 70	108 − 38	84 − 40	139 − 47	59 − 10	867 − 374	32 − 30	643 − 53

Name _____

MULTIPLICATION PRACTICE

	A	B	C	D	E	F	G	H
I	5 ×5	1 ×8	8 ×4	2 ×7	6 ×9	8 ×9	4 ×4	9 ×9
J	9 ×6	4 ×6	7 ×6	3 ×6	12 ×5	5 ×9	7 ×9	1 ×4
K	5 ×0	11 ×6	5 ×4	9 ×7	3 ×8	6 ×2	1 ×9	12 ×8
L	2 ×4	9 ×0	7 ×8	4 ×7	8 ×5	11 ×5	5 ×6	6 ×6
M	8 ×6	3 ×7	5 ×8	10 ×7	7 ×5	3 ×9	9 ×8	2 ×5
N	8 ×1	12 ×6	9 ×3	5 ×1	6 ×4	8 ×0	1 ×8	7 ×3
O	3 ×5	6 ×7	0 ×0	8 ×3	3 ×6	2 ×7	4 ×1	9 ×1
P	6 ×0	4 ×2	7 ×4	3 ×4	4 ×9	11 ×8	6 ×8	5 ×7
Q	3 ×7	9 ×4	10 ×3	2 ×1	8 ×7	5 ×3	1 ×3	8 ×2
R	4 ×8	5 ×2	6 ×3	0 ×6	9 ×2	4 ×3	12 ×4	6 ×1

Skills Practice

143

Practice for TAKS Success

	A	B	C	D
E	$108 \div 9 =$ ____	$28 \div 4 =$ ____	$7\overline{)56}$	$30 \div 6 =$ ____
F	$36 \div 6 =$ ____	$2\overline{)12}$	$54 \div 9 =$ ____	$63 \div 9 =$ ____
G	$2\overline{)10}$	$2 \div 1 =$ ____	$42 \div 6 =$ ____	$8\overline{)32}$
H	$18 \div 2 =$ ____	$8\overline{)8}$	$48 \div 8 =$ ____	$16 \div 8 =$ ____
I	$15 \div 5 =$ ____	$12 \div 4 =$ ____	$2\overline{)18}$	$66 \div 11 =$ ____
J	$9\overline{)72}$	$36 \div 9 =$ ____	$14 \div 2 =$ ____	$12\overline{)72}$
K	$0 \div 3 =$ ____	$11\overline{)88}$	$81 \div 9 =$ ____	$45 \div 5 =$ ____
L	$20 \div 4 =$ ____	$12 \div 3 =$ ____	$3\overline{)9}$	$6 \div 2 =$ ____
M	$1\overline{)4}$	$48 \div 6 =$ ____	$56 \div 7 =$ ____	$0 \div 1 =$ ____
N	$32 \div 4 =$ ____	$54 \div 6 =$ ____	$44 \div 11 =$ ____	$7\overline{)84}$
O	$27 \div 9 =$ ____	$5\overline{)50}$	$18 \div 6 =$ ____	$36 \div 9 =$ ____
P	$1\overline{)8}$	$36 \div 3 =$ ____	$3\overline{)15}$	$0 \div 7 =$ ____
Q	$72 \div 8 =$ ____	$40 \div 8 =$ ____	$24 \div 6 =$ ____	$10\overline{)80}$

Match the definitions with the correct terms. Write term in the spaces below.

Expanded Form	Standard Form	Whole Number	Place Value
Rounding Product	Numerator	Fraction	Digits Compare

1 _____ The answer in a multiplication problem.

2 _____ The symbols 0, 1, 2, 3, 4, 5, 6, 7, 8, and 9.

3 _____ A way to write numbers by using the digits 0–9, with each digit having a place value.

4 _____ A way to write numbers by showing the value of each digit.

5 _____ One of the numbers 0, 1, 2, 3, 4, ... The set of whole numbers goes on without end.

6 _____ Replacing a number with another number that tells about how many or how much.

7 _____ The value of each digit in a number based on the location of the digit.

8 _____ The part of a fraction above the line, which tells how many parts are being counted.

9 _____ A number that names part of a whole or part of a group.

10 _____ To describe whether numbers are equal to, less than, or greater than each other.

Need Help? Check out http://www.harcourtschool.com/hspmath and click on the glossary link or look up the words in the glossary located in the back of the book!

Name the parts of the problem.

addend
sum
addend

_____

$2 + 3 = 5 \longrightarrow$ _____

_____

factor
product
factor

$7 \longrightarrow$ _____

$\times\ 4 \longrightarrow$ _____

_____ $\longleftarrow 28$

Quotient
Dividend
Divisor

_____

6

_____ $\longleftarrow 5\overline{)30}$

_____

Put the number sentences into two fact families.

A fact family is a set of related addition and subtraction, or multiplication and division, number sentences.

$6 + 3 = 9$	$7 \times 3 = 21$	$3 + 6 = 9$	$21 \div 7 = 3$
$3 \times 7 = 21$	$9 - 3 = 6$	$21 \div 3 = 7$	$9 - 6 = 3$

Family 1 _____   Family 2 _____

_____   _____

_____   _____

_____   _____

Sara made this pattern with counters.

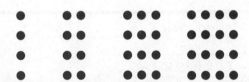

Draw the next figure in Sara's pattern.

Write a rule to describe the number pattern below.

42, 38, 34, 30, 26, 22

Rule: _____

Finish the sentence.

$$103 < 130$$

1 The $<$ means _____.

$$6 \times 3 = 18$$

2 This is a _____.

3 This is an _____.

4 Draw an array for the equation.

$$5 \times 7 = 35$$

5 Shade in all the square numbers on the multiplication table.

+	0	1	2	3	4	5	6	7	8	9	10
0	0	0	0	0	0	0	0	0	0	0	0
1	0	1	2	3	4	5	6	7	8	9	10
2	0	2	4	6	8	10	12	14	16	18	20
3	0	3	6	9	12	15	18	21	24	27	30
4	0	4	8	12	16	20	24	28	32	36	40
5	0	5	10	15	20	25	30	35	40	45	50
6	0	6	12	18	24	30	36	42	48	54	60
7	0	7	14	21	28	35	42	49	56	63	70
8	0	8	16	24	32	40	48	56	64	72	80
9	0	9	18	27	36	45	54	63	72	81	90
10	0	10	20	30	40	50	60	70	80	90	100

● Record the number of equal sides and right angles each figure has. Then draw a figure that is the correct type of triangle in the same row.

Types of triangles	Number of equal sides	Number of right angles	Draw the correct triangle
Equilateral Triangle			
Scalene Triangle			
Isosceles Triangle			
Right Triangle			

Match the figures with the terms.

A. sphere
B. trapezoid
C. right angle
D. cylinder
E. cone
F. pentagon
G. hexagon
H. octagon
I. cube
J. ray

1 _____

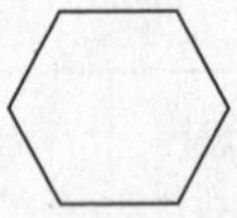

2 _____

3 _____

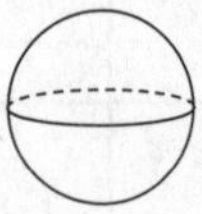

4 _____

5 _____

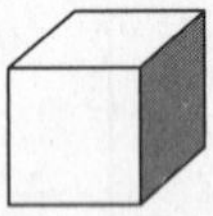

6 _____

7 _____

8 _____

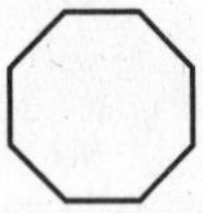

9 _____

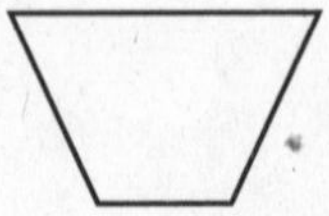

10 _____

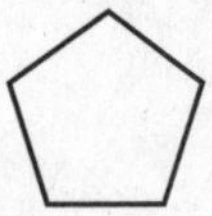

© Harcourt

Name _____ Date _____

MEASUREMENT

Choose the unit of measure for each correct part of the chart below.

| Quart | Yard | Centimeter | Meter | Kilometer | Pound | Milliliter |
| Pint | Gallon | Mile | Pound | Ounce | Cup | Kilogram |

Units of Measurement

Length	1. _____ 2. _____ 3. _____ 4. _____ 5. _____ 
Capacity	6. _____ 7. _____ 8. _____ 9. _____ 10. _____ 
Weight/Mass	11. _____ 12. _____ 13. _____ 

© Harcourt

Vocabulary Review 151 **Practice for TAKS Success**

The distance around a figure is called its perimeter.
Find the perimeter of each figure.

1
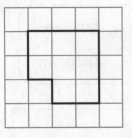

Perimeter = _____ units

2
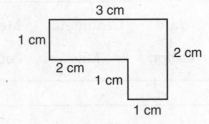

Perimeter = _____ cm

3
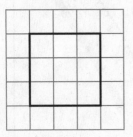

Perimeter = _____ units

4
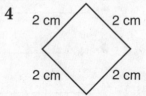

Perimeter = _____ cm

Area is the number of square units needed to cover a flat surface.
Find the area of each figure.

5
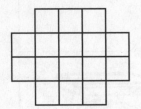

Area = _____ units

6
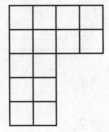

Area = _____ units

Label the bar graph and the scale of each graph. Some words may be used more than once.

> Scale   Horizontal Bar Graph   Vertical Bar Graph

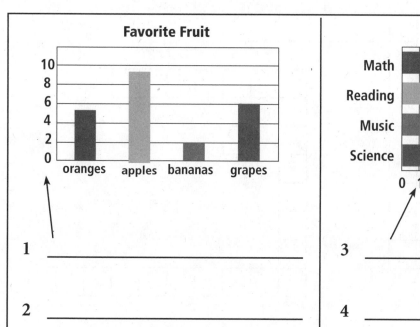

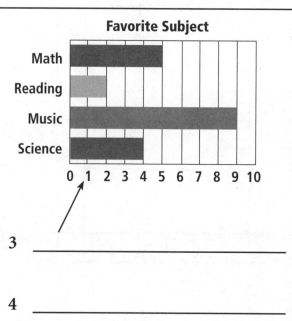

1 _____

2 _____

3 _____

4 _____

An event is **likely** if it has a good chance of happening.
An event is **unlikely** if it does not have a good chance of happening.
Write *likely* or *unlikely*.

5   It is _____ that I will spin white.

It is _____ that I will spin gray.

6    It is _____ that I will pull a white marble.

It is _____ that I will pull a black marble.

Use the clues to complete the puzzle.

	Across		Down
2	An event is _____ if it will never happen.	**1**	An event is _____ if it has a good chance of happening.
4	A possible result of an experiment.	**3**	The chance that a given event will occur.
6	To make a reasonable guess about what will happen.	**5**	Something that happens
7	A test that is done in order to find out something.	**6**	Something that has a chance of happening.
8	An event is _____ if it does not have a good chance of happening.	**9**	An event is _____ if it will always happen.

© Harcourt